BIRDS AND HOW THEY FUNCTION

BIRDS AND HOW THEY FUNCTION

PHILIP S. CALLAHAN

HOLIDAY HOUSE · NEW YORK

Photographs and drawings not otherwise credited are by the author.

Printed in the United States of America

Library of Congress Cataloging in Publication Data

Callahan, Philip S. 1923-
Birds and how they function.

Includes index.
SUMMARY: Discusses the evolution, physiology, and behavior of birds.
1. Birds—Juvenile literature. [1. Birds]
I. Title.
QL676.2.C34 598.2 79-2109
ISBN 0-8234-0363-7

FOR KEN AND IRWIN,

two good friends who, on our hike around the world, always stopped so I could bird-watch wherever we happened to be.

A friend may well be reckoned the Masterpiece of Nature.

EMERSON

CONTENTS

BIRDS AND HOW THEY FUNCTION

1/ BIRDS AND LIFE

What is life? It is the flash of a firefly in the night. It is the breath of a buffalo in wintertime. It is the little shadow which runs across the grass and loses itself in the sunset.

Dying words of Crowfoot, a Blackfoot Indian (1890)

I think I would add to Crowfoot's beautiful definition of life that it is also the scream of the prairie falcon from the canyon wall. For me no living creature seems more alive than the swift, keen bird that courses the sky where prairie meets mountain.

I don't know what there is in me that makes me want to know the deepest inner workings of nature. I am sure, however, that it is my Indian ancestors that placed in my genes their own feelings for nature. Like my Native American forebears I look upon the land as a beautiful woven blanket spread over the globe—a Navajo blanket in which each individual thread of life connects one pattern to another, and in which each pattern is a different aspect of life—one the soil, one the animals and birds, others the insects and plants, and finally in the very center the human race. Of all the creatures woven into the complex pattern of life we are the only ones who seem to have a need to know the inner secrets of nature and to try to discover them by the pursuit of what civilization calls science.

It is unfortunate that in our modern Western society, science has come to mean largely technology, and that modern scientists are considered practitioners of a profession that designs machines to rip apart that living Navajo blanket we call nature.

Science is not technology, nor are scientists technologists.

Technology is the use to which humans put the discoveries of science; and all the scientists I know hope in their hearts that their discoveries will benefit us, and not destroy our country or the world. Even more important, most scientists try to use science in a way that is compatible with the laws of nature.

The Native Americans themselves produced great scientists —or naturalists, if you prefer that word for a biological scientist. Almost every vegetable that Westerners eat was discovered and bred to its present form by one or another tribe of Indians. Of course, we don't know what individual Indian first had the idea of breeding and planting corn or squash, but that does not mean that the Indians were unscientific. They had a tremendous knowledge of plants and animals; and although they lived close to the soil, they looked to the sky and understood the look in the eagle's eye.

The Indian view of life was one with nature, but in spite of that, Indians could be as destructive of nature as any other human society. In Canada it was mainly Indian hunters and trappers that nearly caused the extermination of beavers and other fur-bearing animals in order to meet the Hudson Bay Company's insatiable demands for furs. No human race has a monopoly on greed or goodness.

The Symbol

The Hopi Indians have a symbol for life that resembles a maze puzzle. The symbol is inscribed on an Indian rock near Oraibi, Arizona, that is considered sacred. It depicts life as a child emerging from the womb of the mother earth. The inside lines represent the fetal membrane, and the outside lines the mother's body with the two arms (at bottom) clasping the baby that is emerging from the womb. The straight line at the center is the umbilical cord connecting the child in its fetal form inside the womb to the same child outside after delivery and lying

Tá pú at *("mother and child"), the symbol of life for the Hopi Indians.*

between the two arms.

In Hopi the symbol is called *Tá pú at* ("mother and child"). It is indeed an elegant symbol of life and is applicable to all living creatures, for life starts with the mystery of reproduction, be it a plant or animal. It is only by understanding reproduction that we can even begin to understand how living things function. And just as important to that knowledge is our understanding of death in the life process. Death is part of life, and as Crowfoot pointed out, life is the breath of the buffalo. So also is the death of the piñon jay plucked from the sky by the deadly talons of the prairie falcon. We live to die, and we must die to live—that is the one really great mystery of life.

It has always been popular to look at nature either from the "noble savage" sentimental point of view or contrariwise as a "tooth and claw" matter, depending on whether or not one is a bird-watcher or a hunter. I am both a bird-watcher and a hunter (I hunt with falcons) and I follow both avocations with no feelings of ambiguity or belief that the one contradicts the other. Nature is a system of life in which birth is the beginning and death the ending for an individual. Reproduction is the thread of life that keeps the woven blanket of nature together so that all living things are connected. John Muir, the great California naturalist, said it best. When a friend once asked why he loved the wilderness so much, his reply was, "Because everything is connected to everything else."

Native Americans' love of beauty is seen in their art forms and the poetry of their language. The Hopi word for "bluebird" is *chosovi* and the word for their beautiful turquoise beads is *chosposi*—meaning "bluebird's eye." Those descriptive words tell us of the deep appreciation Indians had for the beauty of birds and yet that did not prevent them from hunting and eating birds of all sorts, including what are generally called songbirds. They used the feathers of birds on their arrows.

The Bow and the Arrow

Saxton Pope, a world authority on bows and arrows, tested the bows of 16 different Indian tribes, including the Blackfoot, Apache, and Navajo. (Bows are tested on a spring scale similar to an angler's outdoor fish scale.) Not one bow pulled over 40 pounds—a flabby pull alongside the modern 60- to 90-pound deer-hunting bow. The simple fact is that most Indian tribes could not have lived by hunting large game. Indians were gatherers of nuts and berries, growers of vegetable crops, and hunters, with their weak bows, of small animals and birds. The Apache bow that Pope tested was made of hickory and pulled less than 28 pounds—a small, short bow suitable only for birds and small animals. Strangely enough, the little Negritos, pygmies of the Philippines and other areas of Asia, hunt with bows far stronger than the bows of most tribes of American Indians. I

A Philippines Negrito with his long bow and six-foot arrow. Note the small bird on the three-pronged tip of the arrow. The wrapping on his arm prevents injury from the snapping bowstring. The G.I. cap and belt are from World War II, during which Negritos helped American soldiers.

once lived with Negritos for a couple of weeks and found that the pull of their extremely long bows (over six feet) was more than 55 pounds. Nevertheless, their main source of protein is insects and birds.

If you look closely at my photograph taken in the jungle of Luzon, the northern island of the Philippines, you will see a small songbird at the tip of the three-pronged arrowhead. My little Negrito friend ate that bird the same day I took the photograph. That should be proof enough of the utility of birds in human diet. Most of my readers probably eat bird meat quite often, for as we all know, chicken is a cheap and delicious protein source. The only difference between other American bird-eaters and Indians is that the Indian probably had a far greater appreciation of the behavior of his feathered brothers than do most scientists or the average city dweller.

The Indian seldom shot at game more than 20 yards away. He was skilled at calling game to within range of his stunning arrows. It was absolutely necessary that he understand perfectly both the behavior and the functioning of his feathered brothers. We no longer hunt songbirds, of course, because there are so many people and so few birds. We find our chickens at the supermarket.

A philosopher once described a great teacher as a person able to speak with simplicity of the profound. The relationship between the human at the center of that great Navajo blanket of life and the pattern of life that surrounds him is indeed profound. A better understanding of birds in our lives than I can give is published in the literature of our Native American brothers. Their writings contain simple explanations of profound subjects.

This book can teach one how science believes that birds function, but it cannot teach one what a bird really is. Knowing how birds function will indeed give one a far greater appreciation of bird life, but if one does not see for himself, then no book can suffice. Go out to the woods to watch and listen; to the desert, perhaps; or even outside your own kitchen door where that little shadow runs across the grass and loses itself in the sunset.

2/HOW BIRDS EVOLVED

Insects are not particularly conspicuous creatures. They dwell in cracks and crevices or hide among the blades of grass. Aside from the beautiful butterflies that flit across fields or woodland clearings, and insects that sting or eat our crops, the average person does not pay much attention to them. Most mammals are even more inconspicuous, for they fear humans and avoid them at all costs. Raccoons and opossums are abundant in the parks and woods of Gainesville, Florida, where I live, but the average resident has never seen a live raccoon or opossum within the city limits.

Birds, on the other hand, are highly conspicuous, and for three very good reasons. They are of course much larger than insects; they fly; fairly many are brightly colored. Because birds can escape by flight, many species are likely to carry on their daily affairs in close proximity to human beings. Also because they fly about, they catch the human eye. Movement is a powerful attention-getter.

It is the beautiful feathers of birds that enable them to fly and have made them so visible and perhaps the most loved of all wild creatures; yet a great many biological scientists agree that birds are descended from reptiles, among the most disliked of all creatures. Why?

To get into this question properly, let me introduce the Nanday conure I own, a species of small parrot related to the much larger macaws of South America. Sam Red-Socks—I named him that because of the bright red feathers on his legs—is from the Mato Grosso of Brazil. His breast and wing coverlets are a beautiful green that is indescribable; his tail has a bluish sheen and his head feathers form a black hood that is cut off just below the edge of the eye. His home is the huge cage that sits on my desk, but when I am around he prefers to sit on top of the cage or on the desk lamp, following the movement of my pencil with his head cocked to one side. Once in a while he pauses to preen his toenails and purr (yes, he does purr) at me. Presumably Sam thinks that I am the greatest thing that ever happened to him and doesn't miss the Mato Grosso one bit. When I look at Sam, my every instinct says that there is no possible way that this beautiful and intriguing creature could have descended directly from reptiles. Biologists, however, have some very convincing reasons for believing that this is so.

In the year 1861, near Solnhofen, Bavaria, Germany, the bony outline of a flying creature called Archaeopteryx (*Archaeopteryx lithographica*) was found imprinted into a block of lithographic limestone—thus the second part of the scientific name. The skeleton is in the British Museum of Natural History in London. A second skeleton of Archaeopteryx was found in 1877 near Eischstadt, Germany, and is in the Berlin Museum. The Bavarian limestone beds date from the Jurassic period of the geological time calendar (see chart).

The bone structure of Archaeopteryx resembles a very small dinosaur with feathers and is about the size of a crow. Ordinarily we think of dinosaurs as great awkward giants and not little graceful birds. The famous English biologist Thomas Huxley suggested over a century ago that the little two-foot-long dinosaur Compsognathus ("elegant jaw") was quite similar to the birdlike Archaeopteryx. Huxleys' theory that birds descended

GEOLOGICAL TIME CALENDAR

ERA	*PERIOD*	*EPOCH*	*YEARS AGO—APPROX. START OF PERIOD*	*ANIMAL LIFE*
CENOZOIC	QUATERNARY	RECENT	10,000 years	Civilization begins
		PLEISTOCENE	2 million	Rise of human species
	TERTIARY	PLIOCENE	12 million	Mammals, birds, and insects dominate the land
		MIOCENE	25 million	
		OLIGOCENE	36 million	
		EOCENE	58 million	
		PALEOCENE	65 million	
ROCKY MOUNTAINS FORMED IN CRETACEOUS PERIOD				
MESOZOIC	CRETACEOUS		135 million	End of dinosaurs; *second great radiation of insects*
	JURASSIC		181 million	First mammals and birds
	TRIASSIC		230 million	First dinosaurs
APPALACHIAN MOUNTAINS FORMED IN PERMIAN PERIOD				
PALEOZOIC	PERMIAN		280 million	Expansion of reptiles; decline of amphibians
	CARBONIFEROUS		345 million	Age of amphibians; first reptiles; *first great radiation of insects*
	DEVONIAN		405 million	Age of fishes; first amphibians and *insects*
	SILURIAN		425 million	Invasion of land by *arthropods*
	ORDOVICIAN		500 million	First vertebrates
	CAMBRIAN		600 million	Age of marine invertebrates

from dinasours was not favored by scientists, but lately some paleontologists are having second thoughts about his idea.

Are Birds Direct Descendants of Dinosaurs?

Archaeopteryx, in which the feathers are easily distinguishable in the imprint, is remarkably reptilian as far as bone structure is concerned. The tail feathers grow in parallel rows from either side of a long bony tail similar to that of a lizard. The beaklike jaws are armed with small teeth. Its brain, which is outlined in calcareous deposits, is clearly reptilian. The pubic (hip) bone is elongated and directed forward like those of birds and the wings are supported by an enlarged breastbone (sternum), as are those of modern birds. There are three digits, fingers with claws, on the wrist of the wings. Some modern birds still retain their wing digits.

Because of these similarities of bone structure, some paleontologists are engaged in a great debate as to whether or not birds are, in reality, modern flying dinosaurs. Many scientists believe that dinosaurs and birds evolved along parallel lines. Parallel evolution means that, although birds and reptiles have similar structures and makeup, they nevertheless developed separately and not in line of descent, and that the resemblance is merely accidental.

Dr. John Ostrom of Peabody Museum, Yale University, has presented very convincing evidence of the descent of the feathered Archaeopteryx from the dinosaur genus Ornitholestes. The skeletons are similar in almost every respect except that Archaeopteryx has feathers and a slightly keeled sternum. The differences are so slight that it was mistaken for a dinosaur for over 20 years before the error was discovered. Because of its feathers, Archaeopteryx must be considered a true bird. It is the first-known creature to occupy the same ecological niche as modern birds.

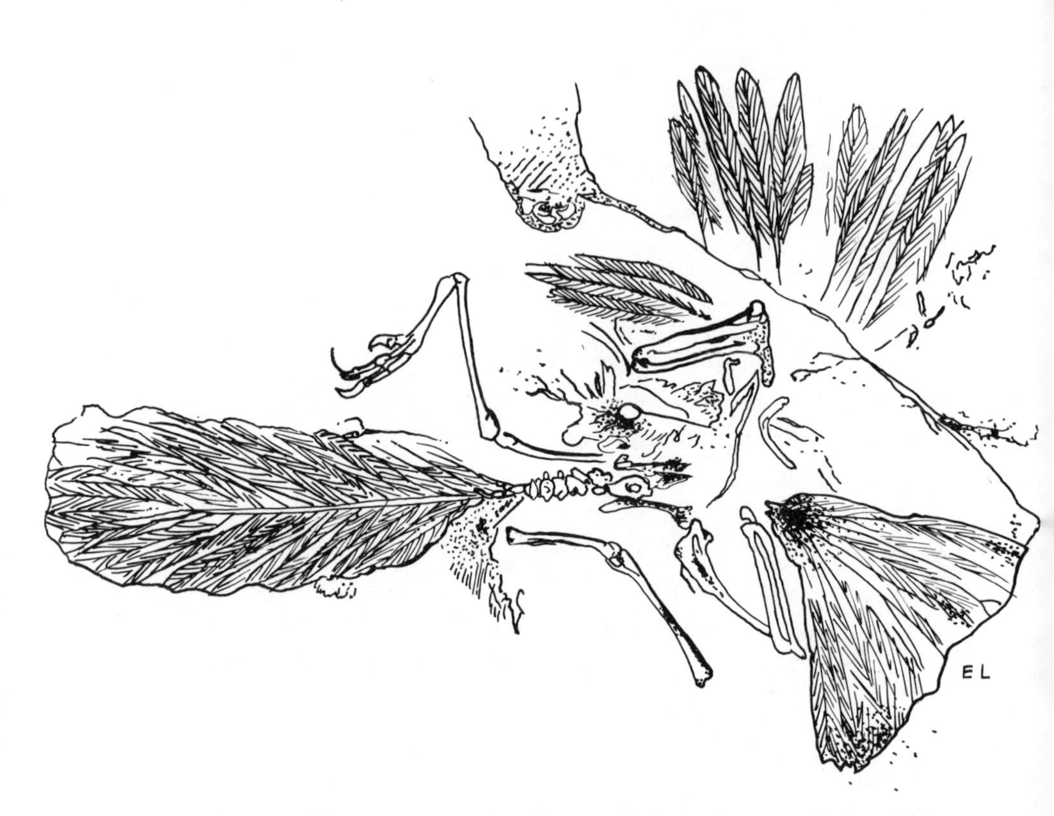

Fossil remains of the earliest known bird, Archaeopteryx. Its bone structure was much like that of a diminutive dinosaur, but it was feathered. Tail feathers grew outward from the sides of a long thin tail.

Modern birds have increased air buoyancy due to their hollow bones—called pneumatized bones. The Archaeopteryx skeleton consists of solid bones, so that the total sum of differences, despite its feathers, between Archaeopteryx and small species of dinosaurs is negligible.

Such slight differences are not likely to indicate parallel evolutionary development. The similarity of birds to dinosaurs leads Dr. Ostrom to suggest that dinosaurs were actually warm-blooded reptiles and not "cold-blooded," as originally believed (and Dr. Robert Bakker, working independently, has said the same). He points out that the breastbone of Archaeopteryx had no large keel, as do our birds, so could not anchor large muscles necessary for flight; therefore the feathers were for gliding and also served as insulation for the body. Since only warm-blooded animals have efficient insulation (hair and feathers), the remains of Archaeopteryx support both the theory of bird descent from dinosaurs and the warm-bloodedness of the dinosaurs. Be that as it may, it is extremely difficult for the nonscientist to connect the giant brontosaurs and tyrannosaurs that once roamed the earth with our beautiful present-day bluebirds and eagles. Strangely enough, the so called featherless flying reptiles, the

In the nineteenth century Thomas Huxley suggested that this two-foot dinosaur, Compsognathus, resembled the bird Archaeopteryx. Though his theory that birds descended from dinosaurs was rejected then, it is considered valid today by many scientists, though not all.
ERIC WATTS

pterodactyls, are not true dinosaurs, so they are probably not even related to birds.

Archaeopteryx Feathers

For many years there has been a debate as to whether or not Archaeopteryx could fly. Some researchers maintain that the feathers evolved as a sort of body net for trapping insects. This seems very unlikely to me.

Two scientists, Drs. Alan Feduccia and Harrison B. Tordoff, maintain that Archaeopteryx was "at least able to glide" and that flapping flight seems a possibility. They base their belief on the asymmetry of the flight (wing) feathers. Most flightless birds have symmetrical feathers—the rachis (shaft) lies along the center of the feather. All flying birds have asymmetrical primaries—the shaft lies close to the leading edge of the primary. Asymmetry assures that the leading edge is thick, stiff, and curved like the front edge of an airplane wing. Asymmetric feathers are clearly visible in the fossil Archaeopteryx.

In my opinion, however, one should distinguish active flight, the ability to take off from the ground, from passive flight, the ability to glide. Both require broad leading-edge airfoils. The heavy, solid bones and lack of a breast keel, for attachment of strong muscles, indicate to me that Archaeopteryx could not rise from the ground. The wing claws indicate a tree-climbing creature with the ability to glide great distances from tree to tree in the manner of flying squirrels. Aeronautical engineers will agree with me that even though gliders must have efficient airfoils, active and passive flight are two different kinds of aerial transport. Archaeopteryx was probably a passive glider.

Birds of the Cretaceous Period

The ability to fly demands a specialized form of an organism. No doubt many different creatures evolved toward flight, but appeared and then became extinct without leaving descendants. The flying lizards, the pterodactyls, are an example. Evolutionists believe that the rapidity with which living organisms were likely to evolve toward flight accounts for the lack of fossil birds. Another reason for their scarcity is no doubt the fragility of bird bones, which are hollow. Only three specimens have been found even of Archaeopteryx, which has solid bones.

If we look at the geological time calendar we see that, since Archaeopteryx is dated in the Jurassic period of the Mesozoic era, true birds and mammals first appeared about 181 million years ago at the middle of the age of dinosaurs, during the Triassic period that immediately follows the Jurassic. The Cretaceous period saw the end of the intriguing dinosaurs and the second great spread of insect forms. From that period, 135 million years ago, only two fossil forms of birds have been discovered. They were found in the Upper Cretaceous limestone of Kansas between 1870 and 1872 and are called Ichthyornis and Hesperornis. Their bone structure is that of true birds; however, Hesperornis has teeth, so it is placed in the "toothed-bird" group called Odontornithes, hence the group name odont, meaning "tooth." They are apparently large birds and were swimmers and divers. Their hind legs were far back and their forelimbs reduced, so they could not fly. They resembled modern-day cormorants, but show reptilian characteristics, especially in the brain.

Members of the Ichthyornis genus were fish-eating birds the size of gulls and were the first-known group to exhibit wings like those of modern birds, capable of flapping flight. Since they were strong flyers they demonstrate that true flight had already evolved over the great Cretaceous sea that 135 million years ago covered what is now central United States.

Birds of the Tertiary Period

After a long gap of geological time a few fossils suggesting rails and cormorants were discovered in New Jersey and dated from the Paleocene period. This brings our evolutionary knowledge of birds to the Eocene period of the Cenozoic era, 58 million years ago. The Eocene is an important period, for many species of reptiles disappeared, but there was an explosion of mammals and bird forms. Over 27 families of our modern groups of birds appeared during this period. This places the evolution of birds as we know them today far back in time compared to humans, since Homo species can claim a mere two million years.

Mammals are also believed to be originally descended from reptiles, but not by way of the dinosaurs. If we look at the traditional though actually hypothetical "tree of life," we see mammals arising from the reptile stem by way of the therapsids, and birds on another branch by way of the thecodonts.

If we are to accept Dr. Ostrom's theory that dinosaurs were warm-blooded and birds descended from them, then we discover a different correct line of descent, as the lineage is by way of the theropods, two-legged carnivorous dinosaurs, through the thecodonts. This of course puts the dinosaurs between the ancestral thecodonts and birds, not in a separate parallel line, and rules out parallel evolution between the two groups. I have never believed the dinosaur group to be "cold-blooded," for some very complex reasons that are difficult to explain. Therefore I accept the descent of birds from the dinosaurs as being closer to the truth.

Mammals that appeared during the Eocene did not at all resemble modern mammal forms, which are only distantly related to Eocene fossil types. Many intermediate forms must have come and gone in the last 58 million years before the present mammalian form appeared. Birds have changed little, however, and fossils of eagles, kites, fish eagles, loons, flamingos, rails, and other waders that have been dated over 58 million years ago greatly resemble their modern descendants. The falconer with a two-million-year-old lineage that holds a golden eagle on his fist has trained a creature with a 58-million-year-old lineage. For this reason among others, let's hope that we can cease gunning eagles into extinction, as we are at present doing despite laws against it.

Vultures, owls, herons, trogans, cuckoos, and shrikes also appeared during the Eocene. Paleontologists say that an explosion of birds occurred during that period. During the Oligocene, 35 million years ago, new species continued to arise at a surprisingly fast rate. Sea birds such as the cormorants, shearwaters, and albatrosses appeared. My friend Sam Red-Socks, the parrot, is from the Oligocene epoch, as are sparrows, turkeys, and pigeons. By the Pliocene, which marks the end of the Cenozoic era, the falcon, lark, stork, swallow, and crow families had appeared.

By the end of the Pleistocene epoch, two million years ago, all modern forms of birds had arisen, and, despite the disappearance of about 30 groups in that period, had reached their maximum diversity of forms. There are about 8600 known species of birds today. Unfortunately, humans today are hastening the extinction of many of them.

3/THOSE MARVELOUS FEATHERS

It requires a good bit of imagination to look at a modern-day alligator or crocodile and believe that feathered birds descended from these unique reptiles, yet Dr. Alick D. Walker of the University of Newcastle-on-Tyne, England, has presented arguments for such a direct descent. He points out that it has long been recognized by paleontologists that birds are more closely related to crocodiles than to other reptiles. The structure of the heart of crocodiles represents an advanced development beyond that of related forms and is on the way to the structure of the bird heart. Dr. Walker maintains that birds and crocodiles are directly descended from early forms of crocodiles. He has made detailed dissections and comparisons of bird skulls with a form of crocodile called Sphenosuchus, from the upper Triassic red beds of South Africa. From the similarities of skull structure, and from embryological and other resemblances between modern bird and crocodile species, he suggests that the two groups are direct crocodile descendants.

Whether or not one accepts the dinosaur or crocodile route of descent for modern birds, the question that remains—since the reptile skin is covered with scales and the bird skin with feathers—is, did feathers evolve from scales?

Dr. Phil Regal of the University of Minnesota, in a well

reasoned paper, has traced a presumed origin of feathers from scales and concludes that scales did change into feathers over the eons.

Birds of course have scaled feet and legs but the reasons for believing feathers evolved directly from scales are quite complex. One of the difficulties inherent in tracing back in time such a gradual change is that feathers are quite delicate and so all that usually remains is impressions of their form in stone.

Scales are not appendages but rather, patterned folds in the epidermis where the scales are formed from the epidermal secretions. Feathers, however, must be considered appendages because they develop, not from patterned folds, but from specialized cells localized on the skin surface. Although they serve to protect the bird from cold and insulate from excessive heat, as does hair on mammals, feathers are much more complex than hair.

Feathers form from a mineralized type of protein and a second chemical called keratin. Keratin is a sulphur-containing protein that is the chemical basis for hair, horn, fingernails, and feathers. It cannot be dissolved in most solvents, and unlike most protein it cannot be digested, even by the enzymes of the intestine. That is why hawks and owls, which swallow hair and feathers from their prey, compact and regurgitate the ball from their crop the following day.

Where Feathers Grow

Feathers grow from a bird's body in lines; these are called feather tracts. Between the tracts are open areas, apteria. Feather tracts are named after the anatomical part of the bird that they lie on, and each tract is divided into regions. The area called the capital tract covers the top of the bird's head; it is subdivided into a frontal region near the beak, a coronal region on top of the head, and occipital region, on the back of the head.

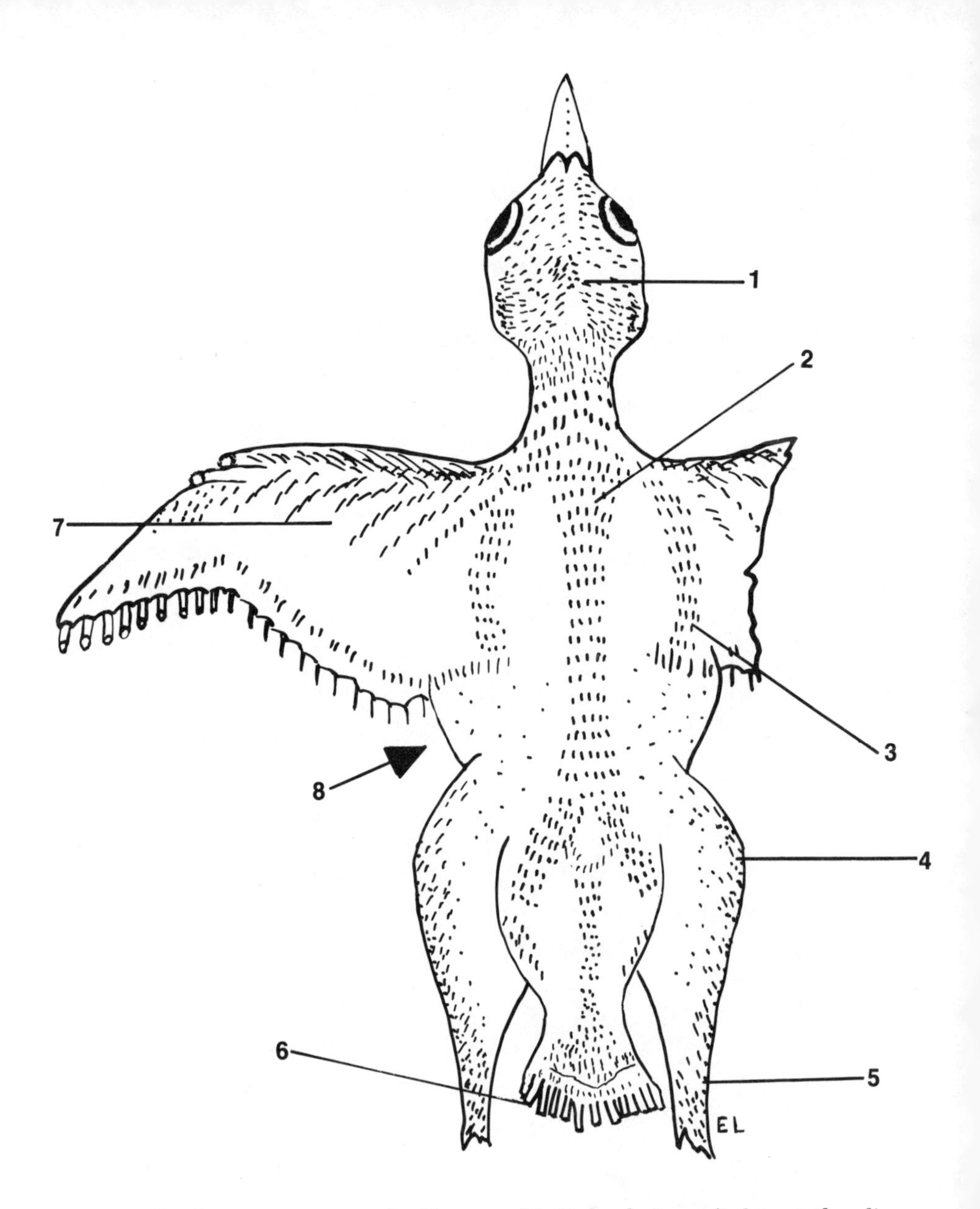

Feather tracts, as seen looking at a bird's back. 1, capital tract (head); 2, spinal tract (neck to rump); 3, humeral tract (shoulder area); 4, femoral tract (thigh); 5, crural tract (leg); 6, caudal tract (tail area, including tail feathers); 7, alar tract (whole wing, including primaries and secondaries); 8, ventral tract (all feathers on underside of bird).

The spinal tract runs along the center of the back from neck to rump; the humeral tract covers the shoulder region; the femoral tract the thigh; and the crural tract the upper legs. The feather tracts serve for insulation and protection of the bird.

There are two major feather tracts of great importance to a flying animal. One, called the alar tract, covers all of the wing area, especially along the leading edge of the wing, and also includes the primary and secondary flight feathers. The second, the caudal tract, covers the tail region and includes the tail feathers. Feathers on the breast and underside of the bird are all grouped together in the ventral tract.

Flight Feathers and Skeletal Support

Tail and wing feathers differ from the tract feathers in that they are attached to bones and evolved to support the bird in flight.

The tail feathers, the rectrices, attach to the last bone of the backbone, the pygostyle. The pygostyle bone is made up of several fused bones that are adapted to hold the tail feathers.

The wing's flight feathers, called the remiges, are attached to the bones of the wings. The outer wing feathers, the primaries, are inserted in the digits—a bird's still-remaining "fingers," in effect—and also to the fused carpal-metacarpal bones (carpometacarpus) at the wing's tip. The carpal bones of the birds are comparable to the human wrist bones.

The inner wing feathers, called secondary feathers, are attached to the ulna, one of the two bones (the ulna and radius) that comprise the middle area of the wing. The ulna-radius joins the large humerus bone that attaches the wing to the bird's shoulder.

Wing joints are highly specialized for flight. The end of the large humerus articulates, or connects by means of a joint, with

few structures that are stronger, lighter, or more flexible. They are one of nature's most marvelous pieces of engineering.

The base of the feather, called the calamus, is hollow and its tip is lodged in the feather follicle on the skin. The feather's shaft, which grows out of the follicle, is called the rachis. The vanes of the feather project out from either side of the rachis and are themselves made up of a series of branches called barbs.

Barbs may be viewed as microscopic feathers, for they in turn are lined with minute filaments called barbules. The edges of barbules form comblike fringes made up of even smaller filaments called barbicels. Some barbicels have at their tip a small hook called a hamulus; this serves to lock the barbicels together in a tight, flexible web. When we pull apart the vanes of a feather it is the hamuli that are separated. When a bird preens its feathers, the movement of the beak along the shaft serves to relock the hamuli.

The feathers that form the bird's outline, tail, and wing feathers are called contour feathers. Fluffy body feathers are semiplumes. Such feathers have a shaft but lack the hamuli that hold the barbules together. Modified feathers called bristles protect the mouth, eyes, and nostrils. Feathers that grow in groups of two to eight around the base of the contour feathers are called filoplumes. Small soft feathers without vanes are the down feathers that we stuff pillows with. A down feather has no shaft or vanes, and the barbs fan out from the tip of the calamus. Down cannot be seen in mature birds because it is

Top: *secondary wing feather of a hawk, as seen by a scanning electron microscope at about 100X; A, one of the barbs composing the vanes on each side of the large central shaft, or rachis; B, barbules, or branches of the barbs.* Bottom: *A, barbules seen in detail at about 500X; note the fringe of barbicels along the edge; B, hamulus, one of the hooks at the tips of the barbicels. Hamuli and barbicels lock the barbules together, and these in turn hold the barbs together to form the feather's vane.* THELMA CARLYSLE AND AUTHOR, USDA

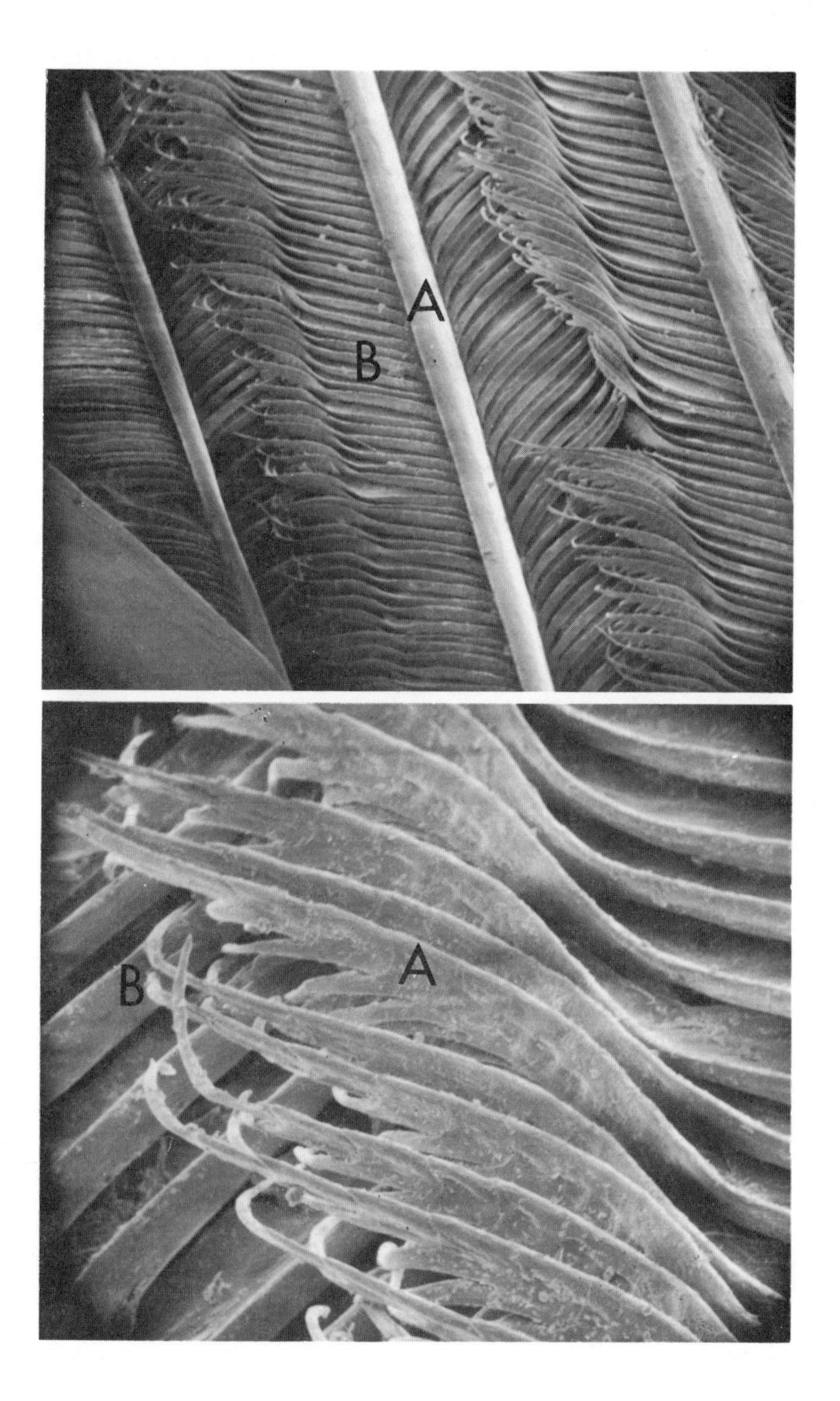
A
B
B
A

concealed under the contour feathers. Some young birds, such as birds of prey, are completely covered with down, which is molted as the bird matures.

The number of feathers on birds depends on the species. Generally speaking, as would be expected, the smaller the species the fewer the feathers. Numbers of feathers vary from 1000 to 3000 among hummingbirds and small warblers, through 6000 in gulls up to as many as 25,000 in large swans.

Pigment and Color

Birds live in a world of color. Most mammals communicate largely by scent but birds depend mainly on the form and color of their plumage. Other than butterflies, birds are often considered the most beautiful of all creatures. It is the color of plumage that so attracts birds to the visual senses of humans. The colors of feathers are dependent on two main mechanisms, one chemical and one physical.

Absorption and reflection of color are characteristic phenomena of pigments. These chemical compounds absorb a part of the white light from the sun, filter it, and—for green, let's say—reflect one narrow set of wavelengths of the total color spectrum. The eyes of birds and mammals are sensitive to this span of colors, made up of violet, blue, green, yellow, orange, and red. There are of course pigments that reflect all manner of hues between these six basic colors.

Pigments called melanins are responsible for the grays, browns, black, and yellowish-brown color of many birds. The chemical molecule responsible for yellow and reds belong to a group of pigments called carotenoids. While melanins are produced by special body cells, the carotenoids are not self-produced and must be taken in through the bird's food. That is why the beautiful pink flamingos fade in zoos unless they are fed certain crustaceans that are rich in carotenoid pigments.

The melanin granules migrate to the barbs and barbules of the feathers, where they are laid down in layers between the thin layers of keratin that form the feather. The carotenoids are contained in fatty inclusions in the forming feather but are eventually dissolved and concentrated in the keratin itself instead of separately between the layers.

Blue feather colors in birds are produced by highly specialized feathers that have barbs with a peculiar physical structure called "Tyndall blue," after an outstanding physicist who explored light and heat. These specialized barbs have an outer layer of refractive keratin beneath which are lines of peculiar "box cells." The walls of these boxes are about two to four micrometers thick and surround 10 to 30 hollow spheres called microvacuoles, each 0.25 micrometers in diameter. Each vacuole is separated by keratin walls less than 0.15 micrometers thick. Each cell box resembles a sponge that physically "traps" the red colors and separately reflects out beautiful blue colors. The size of the little microvacuoles in the box cells determines the hue of blue that is reflected from the barbs of the feather.

A second type of physical color is due to what physicists call optical interference. Such color depends on the layered or the corrugated structure of barbules that are considerably modified in form in such a way as to increase the reflecting area. Interference colors are produced by the strengthening (addition) or weakening (subtraction) of certain wavelengths of color due to the one color's interference with another. The phenomenon allows for a considerable number of brilliant hues to be reflected from the feathers. Usually the brilliance of the color depends on the angle from which the plumage is seen. That is why birds such as the purple grackle appear black when seen from one direction and suddenly become a brilliant metallic purple when caught by the eye at a different angle.

Some birds such as hawks, owls, and parrots appear to have a flat matte surface. This is mainly due to powdery particles about one micrometer in diameter that cover the plumage and comes from the disintegration of special powder-producing down feathers.

4/ GETTING ALOFT AND STAYING ALOFT

Feathers probably did evolve from scales but in the process they developed into extremely strong and lightweight structures. Try to pull a feather apart and you will soon realize that they are almost indestructible. Birds expend a considerable amount of energy on the care and cleanliness of their feathers. Birds with disordered feathers cannot fly well, so they spend much time preening and grooming them. They preen by taking each wing and tail feather separately in the beak and, starting at the base, run the bill along the entire length of the feather. This straightens the tiny barbs and barbules, rejoining these hooks along the length of the vanes.

Almost all birds have small oil glands just in front of and above the base of the tail feathers. They reach around with their beak and nibble the oil gland so that the bill becomes oil-coated. As they preen, the film of oil is spread along the vanes of the feathers and waterproofs them. It is especially necessary that water birds keep their feathers well oiled; otherwise they would become waterlogged and could not fly. Waterlogged feathers mean the death of the bird.

Birds that are healthy shake and ruffle out their feathers and get rid of any small, loose body feathers. Falconers call preening "dressing" the feathers, and shaking the body feathers

The way that birds drop feathers in molting varies according to species. This little blue heron—a species that is white when immature but turns to blue as an adult—shows how big birds such as herons and hawks lose their feathers in pairs when they molt; pairs of the young white feathers remain among the color-changed primaries and secondaries.

"rousing." In one of the earliest books on falconry, titled *Boke* [book] *of St. Albans* (1486) we read that "rowse" is "when a hawk lifteth herself up and shaketh herself." If you have a pet bird and notice it rousing and preening you know it is in good health; if on the other hand it ruffles out its feathers and leaves them puffed out, it is probably sick and has a temperature. A healthy bird always pulls its feathers back in tight against the body after preening and rousing.

Molting

Molting is under the control of the endocrine system and is an extremely complex process. Feathers take very hard usage in the process of the bird's activity, especially during flight, and must be replaced. Strangely enough, feathers form themselves almost like a long chain of connected sausages. Each feather begins as a small feather bud in a pit called the feather follicle. The follicle contains the new feather shaft. The old feather shaft is connected to the new one but is demarked from it by a slight constriction. As the new feather pushes out from the bird it becomes increasingly strong through keratinization, until eventually the old worn feather breaks off at the constriction. The new one is then free to expand outward from the enveloping sheath.

Molting is synchronized with the breeding season. Most birds molt only once a year, but in a number of species there are two molts, one before and one after breeding; the latter is called the postnuptial molt. Obviously, most species of birds do not drop all their feathers at once. You will note in the photograph of the little blue heron that there are two white primaries on each wing and two secondaries. The little blue heron is pure white the first year of its life. Fledgling birds that are not sexually mature are often colored differently from the adults. Large birds, such as herons and birds of prey, lose their feathers in pairs as they molt. Most birds drop only a few feathers at a time over an extended period; thus their flying ability is not impaired. The pattern of dropping feathers varies from species to species.

There are some species that molt all of their feathers at once, especially in the waterfowl groups. Wild ducks in North America nest in the far north and after breeding retire to isolated lakes and other bodies of water in the tundra where they are fairly well isolated from predators. After getting their new flight feathers they head south in August. Wild ducks from central Europe, such as pintails, gadwalls, and shovelers, congregate at the mouth of the Volga in Russia to molt in comparative safety. There are high isolated lakes in the Himalayan mountains of Tibet that serve as refuges from predators for ducks migrating from northern Siberia to southern Asia. The flight to molt and regrow feathers, such as occurs in Tibet, is called a molt migration.

In a marvelous book edited by Roger Durman, *Bird Observatories in Britain and Ireland,* there are listed the best spots for observing bird migrations across those lands. Most birds that migrate across Ireland and the British Isles follow the coast of Europe or cross France to North Africa and have already completed their postnuptial molt. It is essential that their wings and tails be fully feathered, for the aerodynamics of bird flight depends on the stability of those appendages.

The Aerodynamics of Flapping Flight

A bird wing functions in the same manner as an airplane wing. Such a lift system, called an airfoil, is designed to have a broad leading edge and a thin trailing edge. The secondary feathers of birds are attached parallel to the bird's body and represent the airfoil, or lift portion of the wing. Resistance to the forward movement of an airborne object is called drag, and

THELMA CARLYSLE

An osprey lands on its nest on a telephone pole. Note the back-thrust of the wing, and the slot action of the primaries and secondaries.

the force that opposes the weight of the object is called lift.

Lift is generated when the flow of air above and below the wing is sufficient to create a partial vacuum above the secondaries. The broad leading edge of the wing causes the air striking the wing to flow upward across the curved top. Since it has to travel farther over the curved top than across the bottom, the air velocity across the top increases. This greater velocity creates a partial vacuum (a drop in pressure) above the top surface. The air below the wing has a shorter distance to travel, so it moves much more slowly. This produces an increase of pressure beneath the wing and forces the wing upward into the partial vacuum above.

In order to overcome gravity, the bird has to move forward fast enough to generate lift. The primaries create forward motion, since they are so arranged that they strike the air at right angles to the bird's body. They function as vibrating propellers and the same lift-airfoil principle is involved as in the case of the secondaries.

As each separate primary feather is forced downward and forward, it serves to function as a miniature airfoil. The downward force pulls the bird ahead, as there is more pressure behind each feather than in front. Collectively they pull the bird ahead as the wing is forced forward into the leading partial vacuum. The flexible feathers can automatically change pitch (twist) according to their position in flight. During the backward upstroke, the feathers twist and push backward against the air; this flexible motion forces the bird forward. Thus the forward motion, which overcomes drag, is generated on the upstroke and downstroke.

There are several types of flapping flight but for comparison we will examine the two main types which are dependent on the wing shape.

The first type we may call broad-winged and is found in woodpeckers, crows, game birds, most small perching birds, eagles, and buteo types of hawks. This kind of wing is characterized by its great width and short length. Because broad wings have large lifting power, such birds are quite maneuverable, and fly slowly with fairly rapid wingbeat. Broad-winged birds can take off easily but are not capable of great endurance or long flights. Forest-dwelling birds such as the Cooper's hawk are excellent at avoiding obstacles. Their short, rounded wings are well suited for dodging among trees, as is the extremely long tail, which serves as a fast-turn rudder. Night predators such as the owls are similarly constructed.

The second type of wing is long and narrow and is found in falcons, swallows, terns, and most sandpipers and other shore birds; and also in waterfowl. Such birds fly fast with rapid wing-

TABLE 1

Wingbeat frequency of several bird species, as determined by radar

SPECIES	FREQUENCY IN BEATS PER SECOND
ruby-throated hummingbird	59.4
house sparrow	19.5
field sparrow	17.2
black-and-white warbler	16.8
Carolina wren	14.6
song sparrow	14.3
ovenbird	14.7
towhee	13.3
cardinal	10.2
eastern meadowlark	9.8
robin	9.0
blue jay	6.1

An American egret, one of the herons, in flight. Such slowly flapping, broad-winged birds have a medium wing load.

beat but cannot remain airborn at very low speeds. Whether or not they are maneuverable depends on special features of the wing. Falcons are highly maneuverable due to their ability to spread or fold their wings while in flight, whereas ducks, which have heavy bodies in relation to their small wings, are not maneuverable at all. Speed and maneuverability depend on a factor called wing loading.

Wing Loading

Wing loading is the weight of the bird (measure of lift) for each unit surface of wing. It is obtained by dividing the weight of the bird in grams into the wing area in square centimeters. (Table 2). A modern jet transport may have a wing loading of 20 to 30 pounds per square foot.

From Table 2 we see that the broad-winged, slow-flapping heron has a medium wing load of 2.5 whereas the heavy-bodied peregrine falcon, with less than half the wing area of the heron, has a loading of only 1.1. Note that although the mallard duck is exactly the same weight as the much larger heron, it has a very small wing area and a load number of only 0.7. The tiny hummingbird with a large wing area and light body of only 3 grams has a high load number of 4.1.

Wing loading per area of wing determines the flight characteristics of bird species. A high wing-load number means less weight to support per unit area of wing and is required for soaring, long-distance flight over vast stretches of country.

Powerful wing muscles and low wing-load number (high wing loading) are characteristics of fast-flying birds like falcons and ducks. The table shows which species are likely to be gliders and soarers and which high-speed flappers.

TABLE 2

Wing Loading

SPECIES	WEIGHT IN GRAMS	WING AREA IN SQUARE CENTIMETERS	WING LOAD PER UNIT AREA
heron	1408	3590	2.5
mallard duck	1408	1029	0.7
peregrine falcon	1222	1342	1.1
crow	470	1058	2.2
mourning dove	130	357	2.7
song sparrow	22	86	3.9
chimney swift	17.3	104	6.0
ruby-throated hummingbird	3.0	12.4	4.1

Why Birds Migrate in Formation

It is believed that long-distance migratory birds, such as geese, adopt the V formation in order to reduce the amount of energy they use in flight. According to certain theoretical calculations, 25 birds in V formation attain 70 per cent increase in flight distance beyond that of a bird traveling alone. The advantage increases with a stiff tail wind. The extra range of formation flight is dependent on the vertical wash behind each wing in the group. There is a very strong upwash beyond a

White ibis rising from feeding grounds. They rise in pairs (those at right have not yet made their final formation) and fly wingtip to wingtip, which helps by providing strong upwash.

forward-moving wing tip. This creates a favorable forward interference that is used by other birds flying abreast in side-by-side formation. The effect is the same as that for a hawk or vulture flying in a thermal upcurrent where less total lift is needed because of the rising hot air. The less the space between one bird's and another bird's wing the less the drag factor. Birds in a V formation adjust their forward motion so that each individual flies at its maximum lift-to-drag ratio.

The above explanation applies to line-abreast flight but only the birds in the center reach maximum efficiency, because they are in the upwash of birds on either side. In the true V formation the center bird flies in a weaker upwash thrown up on both sides from the birds flying behind. Those at the rear of the formation fly in a favorable upwash on only one side from the birds directly ahead; however, they receive the benefit of a total upwash from all of the birds ahead of them. In this type of V formation the drag savings are evenly distributed among all of the birds. The close cooperation of birds in the V formation not only increases their own flight efficiency but also provides us with that breathtaking symbol of Indian summer—long V's of geese against autumn skies.

5/ GETTING AROUND ON THE GROUND

Just as wings fit the bird for its aerial habits, the evolved shape of the feet fit its ground niche and way of life. Birds' legs may be modified considerably depending on whether the bird hops, walks, climbs, or swings. Some birds such as chimney swifts are almost completely aerial in habit and walk with great difficulty on their short stubby legs; others, such as the roadrunner, have evolved extremely efficient legs for a high-speed chase after their prey on the ground. As mentioned in the last chapter, the forelimb (wing) is supported by the shoulder girdle and sternum. The legs are supported by the spinal column and pelvic girdle. The former is a horizontal support system and the latter a vertical support system. The two are connected by the rib cage.

The legs connect to the pelvic girdle. The three bones of the U-shaped pelvic girdle, called the ilium, ischium, and pubis, are fused into a strong single structure. The U is upside down and the pubic bones, which are the ventral (underside) bones, elongate into two slender rods that are free at their ends and allow the eggs to pass through the pelvic girdle. The thigh bone, or femur, connects into a socket where the fused ilium and ischium join. The bird thigh is completely concealed under the contour feathers of the body. For that reason the first visible parts in a bird's leg are not the thigh and knee joints but the feathered leg portion below the thigh.

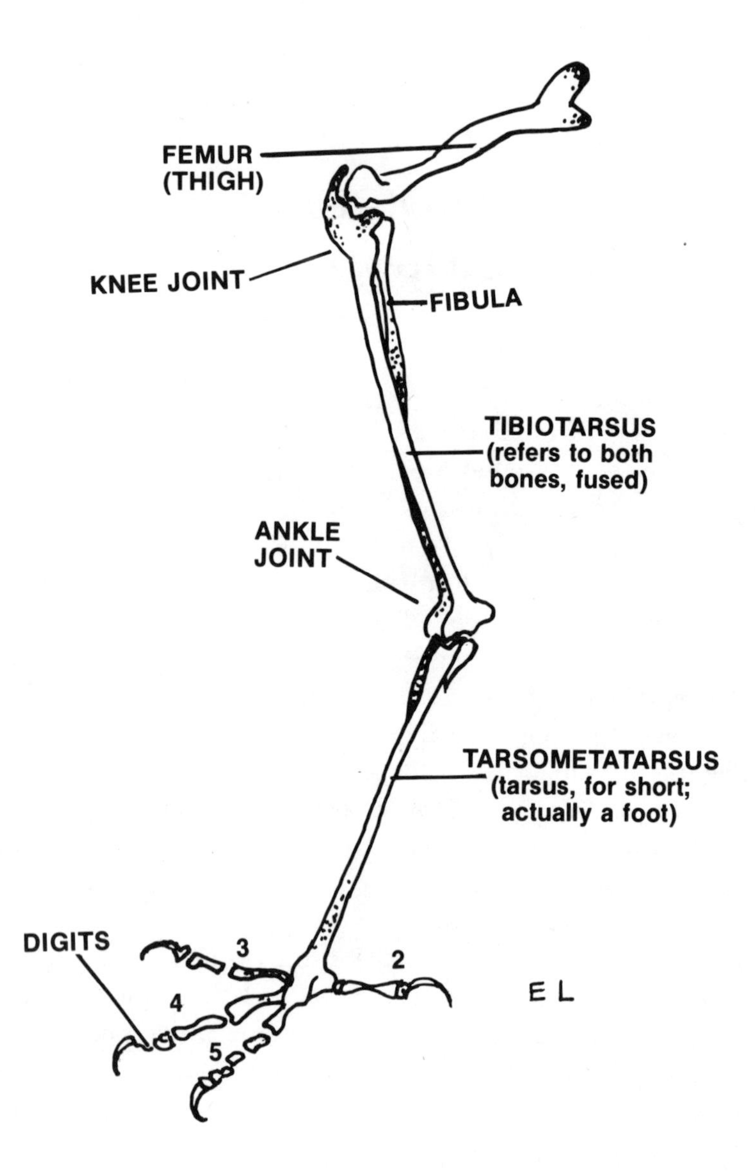

The leg bones of a bird. The digits, or toes, contain 2, 3, 4, and 5 bones respectively.

People often confuse the ankle joint of a bird with the knee joint of other vertebrates. The bird knee is concealed at the tip of the covered thigh. If you look closely you can understand that the first visible joint of the bird's leg cannot be a knee since it bends backward instead of forward like a real knee. The top visible portion of a bird's leg is supported by the fused tibiotarsal bone. The tarsal bones of humans support the ankle. The lower portion of the bird's leg is supported by the remaining tarsal bones, which are fused with the metatarsal bones to form the tarsometatarsus (called tarsus for short). This lower portion of the bird's leg is scaled and is also the most easily visible portion. This long, lower scaled portion of the leg is really an elongated foot.

The Feet

The tarsal bone connects to the four digits that form the toes (claws) and which are composed of two, three, four, and five bones respectively. Usually the toes are arranged three forward and one backward; in this case the organ is called an anisodactyl foot. Such a foot makes a formidable set of pincers. The tendons of the leg and foot are modified so that all the toes close simultaneously and can lock on a branch like a pair of pliers. Some birds, such as owls, cuckoos, parrots, and woodpeckers, have two toes forward and two backward—a zygodactyl foot. Most American plovers and the auks and guillemots have tridactyl feet—three toes only, going forward. Among a number of birds there are exceptions to the general rule; for example, there is a three-toed woodpecker.

Perching and Climbing

Songbirds are in the order called Passeriformes, and include the crows, jays, thrushes, warblers, sparrows, and most

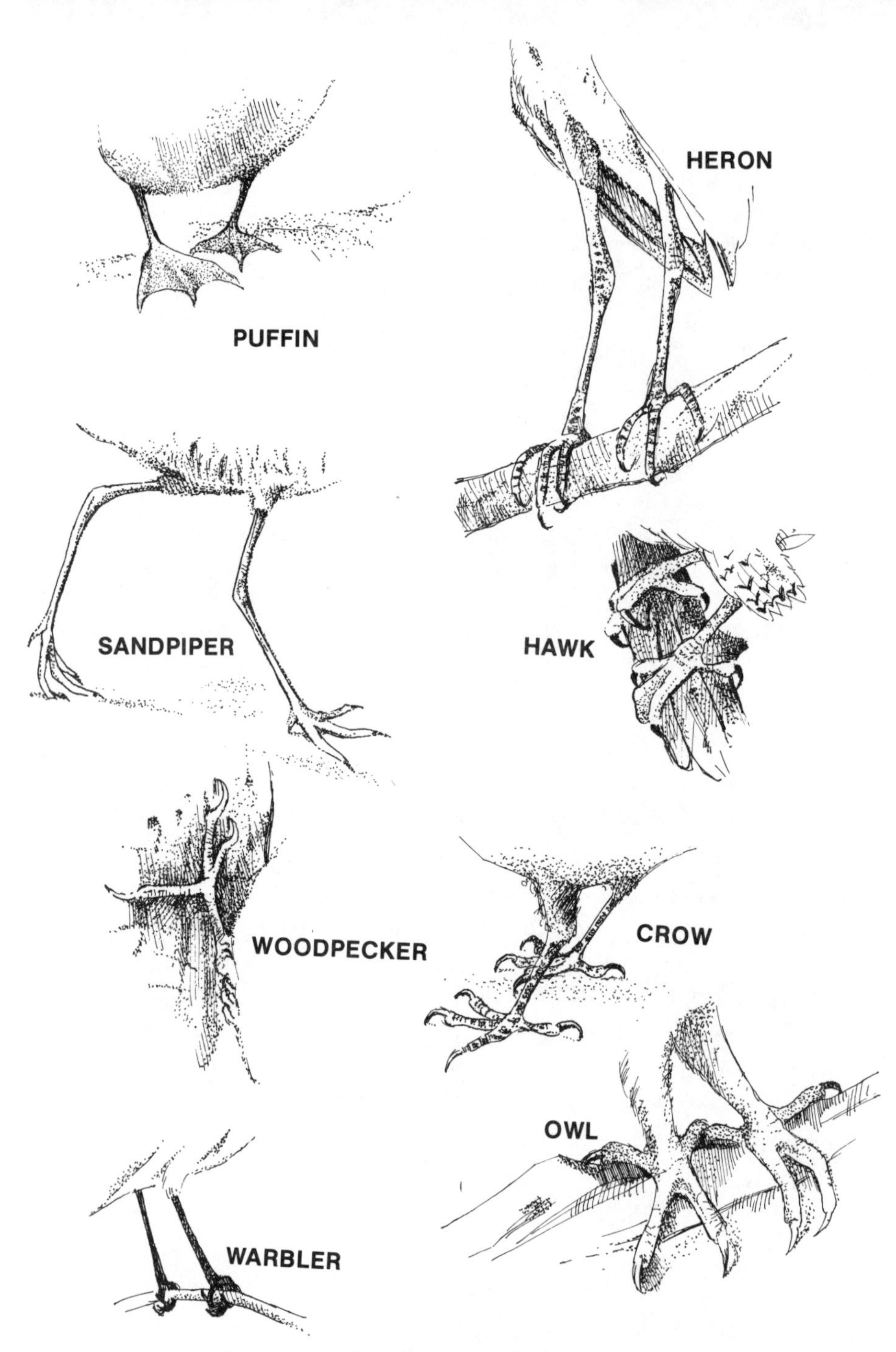

The feet and legs of various birds. ERIC WATTS

common land-dwelling birds; they are called perching birds. They have a short tarsus and pincerlike toes well adapted for gripping a branch. Perching birds of the Passeriformes tend to be tree-dwellers and primarily insect- and seed-eaters. That is not to imply that even web-footed birds, the swimmers, do not perch at times. At almost any season the awkward-looking brown pelicans can be seen perched on the railings of fishermen's wharfs along the beaches of Florida.

Walking and Hopping

Birds that are good at walking or running, such as the roadrunner of the West or the savannah hawk of South America, tend to have long legs. The long legs of the savannah hawk not only fit it for good locomotion but also for wading in marshy areas. They also give the hawk height to spot small prey while on the ground level. A bird of the savannah and open marshy areas, it feeds mainly on frogs, lizards, and insects. When hunting, it hops down from a perch to the ground and peers around from the vantage point of its stiltlike legs.

Birds that have evolved mainly for ground movement run or walk by moving their legs alternately. Walking is the gait of the ground-nesting game birds such as quail and also of larks, magpies, crows, and starlings, all of which tend to feed on the ground. Walking birds usually have long narrow bodies and move horizontally to the ground. Their toes are long, giving flat support. Large birds such as hawks and herons tend to walk, since hopping would involve considerable energy for a heavy bird.

Small perching birds usually hop, since it allows for fast movement and is compatible with their light weight. Most hopping birds are tree-dwellers and spend comparatively little time on the ground. The hopping gait consists of a series of hops separated by short pauses.

Birds, such as this limpkin, that feed in wet places usually have long legs and toes.

Birds that find their food in wet or swampy places tend to have relatively long legs and long toes. Some, such as the Florida gallinule, have extremely long toes, which aid walking on floating lily pads and water hyacinths. The long toes of bitterns are especially suited for "reed-walking." Such toes give excellent support without the added weight of webbing between the digits.

Birds that have evolved as climbers include the parrots, woodpeckers, and cuckoos. The lateral (outside) fourth toe of the climbers is turned backward. Climbing birds also have highly curved claws ideal for hooking into and piercing the bark of trees.

Swimming

The feet of swimming birds are webbed and placed far back on their bodies. Swimmers, and especially divers, must thrust against the water, and the posterior adaptation is extremely efficient for water propulsion. The web stretches between the toes and functions in the manner of a paddle. In ducks, geese, and sea gulls the web stretches between the forward three toes. In many sea birds such as cormorants, gannets, and tropical marine birds the web includes also the hind toe.

More will be said about wing, leg, and foot modifications in relation to food-gathering habits of the various types of bird appendages. The search for food is directly dependent on these.

6/ IN SEARCH OF FOOD

A bird's digestive tract begins with its beak. As in the case of all vertebrates, the mouth is the entrance to the digestive tract but the method by which the food is processed before swallowing depends on the food habits of the animal.

The jaws of mammals are modified for grinding or tearing food, and teeth have evolved according to whether or not the mammal grinds its food, as do vegetarians, or tears and grinds, as do most of the carnivores. The jaws of birds have evolved into beaks, which are efficient and powerful tools for collecting food. However, the beak is not used to prepare food in the sense that it is ground up to be swallowed. Food processing by the beak is usually limited either to tearing, in the case of birds of prey, or extracting seeds or nuts from their hard outer covering. The beak is a light, efficient gathering tool and its shape depends on the habits and ecological niche of the bird.

A beak is composed of horny sheaths called rhamphotheca that cover the jawbones. The toothless jaws are reinforced and very strong along the edges, which are called tomia. The bones that support the beak are light, and, as in the case of snakes, joined quite loosely to the base, where they join the skull. This allows for considerable movement around the mouth.

In some birds the jaws are also hinged on top. There is a

Beaks come in a great variety. ERIC WATTS

hinge of strong cartilage that permits the beak of the parrots, woodpeckers, and woodcocks to move upward.

In most birds the horny sheath extends right to the base of the bill but in others the base is covered with a soft skin area, usually colored yellow, called the cere. The nostrils open on the cere in hawks, pigeons, and parrots.

The bird's light beak evolved along with a high metabolic rate and development of flight; however, it has several functions besides taking up food. In each species it is always the minimum size required for its food habits. It also enables a bird to pick up things, is used as a defensive weapon in a fight, and helps greatly in nest-building. The beak has a lighter jaw skeleton and less bulky muscles in its jaws than mammals or other vertebrates have.

A raptor, or bird of prey, has a hooked beak that is useful for dealing with its kind of food, as are the powerful claws. This is a golden eagle.

Types of Beaks

A bird's beak is mainly a food-gathering tool; as such, it has the necessary shape to gather the kind of food a species needs, in the sort of terrain or surface where the food lies (or flies). There are a great many types of beaks, too many to list here. A survey might start with the little, delicate beak of a wood warbler, just right for catching insects, and end with the long, strong pelican's beak with its gular pouch hanging below that stores fish.

Raptors, or birds of prey, have beaks that are hooked, which means they are well adapted for ripping apart pieces of meat. Though many people believe that falcons, hawks, and eagles use their beaks for striking out in self-defense, this is not so; a cornered bird of prey uses its talons, spread wide, as its defense. The falcon bill has a notch at its tip. A peregrine falcon, which is one of the swiftest flying birds known, swoops down on its prey from well above it in the open air. A steep dive—called the stoop—brings it down with open claws ready for the kill.

Only recently has a misunderstanding about the nature of the stoop been cleared up. For many centuries observers thought that a falcon hit the prey with its claw clenched like a fist. Slow-motion movies, however, have now demonstrated that the claws open up at the last second, just at the instant of the strike, and the hind talons rip into the prey's back. If this is not enough to kill the prey bird, the falcon takes it swiftly to the ground and breaks the neck vertebrae with the notch in its beak. Raptors generally take a rest after they have made a kill, and pluck the prey before eating it.

The parrot is another type of bird with a hooked beak. The family includes cockatoos, lories, parakeets, lovebirds, and macaws. Such a bird has very powerful muscles controlling the beak and can actually gnaw through a thick piece of wood. The strong, hooked bill makes an efficient seed-cracking tool, and is

The long neck of a heron allows it to thrust its daggerlike beak down between thick reeds or lily pads, or into the water, to grasp frogs and other small marsh creatures.

also helpful as a sort of alpenstock in climbing rough surfaces, such as tree bark or the bars and crosspieces of a cage.

One of my favorite birds is the magpie. During my life I have several times had one or two around for pets. It has always been a persecuted bird, except in China, where it is considered a bird of good omen. In the West it is considered a wicked bird and legend has it that its so-called evil temperament stems from its refusal to go into the ark with Noah. Perhaps that is why I am so fond of magpies—they are so independent; even as pets they do as they please. They are also omnivorous, eating almost anything, including insects, fresh meat, carrion, or seed. Their beak is an efficient tool for food-gathering.

Many Adaptations

Magpies, crows, and jays, like parrots, have powerful and sturdily built beaks. However, they have greater length and are not adapted specifically for cracking seeds open. A magpie's bill has varied functions; it is used to pry about under the plant litter lying on the ground, to dig into the soil itself, and to hammer on dead wood or on acorns and other seeds to open them. Magpies, like the squirrels, put acorns away in storage. In the last two months of the year the yellow-billed magpie spends up to 20 per cent of its time transporting and storing these seeds of the oak for later eating. As in the case of squirrels, it is probable that this habit helps to spread new oaks in an area, since they often bury the acorns, driving them in with their beaks.

The heron, which lives on fish, frogs, crayfish, salamanders, and other small water animals, has a tool well adapted for feeding; it is long and pointed, serving as a spear as well as forceps. In the case of this bird, the bill *is* used for defense; a cornered or wounded heron (or its close relative the bittern) will strike out at the eye of a human or other threatening creature. Kings of the Middle Ages used to hunt herons, depending on peregrine falcons to make the kill, but often such falcons became victims themselves at the expense of a heron's sharp-pointed, slashing "sabers."

Snipes have long, slender beaks, ideal for poking into mud, bogs, and tundra. Among the animals it pulls out of muddy places are wireworms (not worms, but the larvae of click beetles), cutworms (larvae of certain moths), crane-fly grubs, snails, crustaceans, and worms.

The woodpecker beak is not only long but also chisel-sharp, and hacks openings through solid wood to find edible grubs. Once the larva is exposed, it uses its tongue, not its bill, to pull it out of its formerly safe chamber; this organ has developed through evolution to a length well beyond the bill's length and has the flexibility needed to yank out the food. It is attached to the hyoid apparatus, which is a set of supporting bones that are attached to the base of the tongue and extend in a loop partly around the skull.

The red crossbill has a strange-looking bill that is crossed at its tip. This species eats the seeds from pine cones. Though at first glance such a beak seems a poor tool, the crossbill efficiently wrenches open the cone scales and then picks up the seeds with its tongue.

The puffin has a strange, marvelous beak which makes it a curious-looking creature. The strange appearance of the bird is due primarily to its large, brightly colored bill. It is flattened laterally and banded with red, blue, and yellow, and embossed with horny enlargements during the mating season. These ridges, called excrescences (from "to grow out" in Latin), are sloughed off after the nesting season. The puffin is one of the few birds that displays its wedding garment on its beak.

The Digestive Tract

The bird's tongue comes in as many forms as the beak. A parrot's tongue is broad and fleshy; thus the increased ability to

The digestive tract of birds has many functions. One example is regurgitation of partly digested food into the mouth of the young, as shown by this female least bittern.

imitate the speech of humans, who also have a broad, fleshy tongue. In some fish-eating birds the tongue is hardened with keratin and covered with spikes that point backward, preventing the fish from slipping forward. In most birds such as crows, songbirds, and others the tongue is sharply pointed and keratinized at the tip. In ducks the base of the tongue has small ridges called lamellae which meet other lamellae from inside the bill to form a filter system. Hummingbirds and the various honeyeaters of tropical climates have tongues in the shape of a tube that is efficient for sucking nectar from flowers.

The salivary glands, as would be expected, are highly developed in woodpeckers and most insect-eating species. They coat the tongue with a sticky substance that holds captured insects. In contrast, most herons and other aquatic birds that swallow their prey whole have reduced salivary glands.

The throat, called the esophagus, is wide in most birds, especially those such as the herons and birds of prey that swallow huge chunks of meat or whole prey. The crop, which is merely a widening of the esophagus, is a storage area and not one of the main organs of digestion.

In birds of prey the crop is well developed for holding the meat as the prey is torn apart. After digestion is accomplished, a neatly compacted pellet (called a casting by falconers) of indigestible hair or feathers is stored in the crop and eventually regurgitated. Owls, which cannot digest bones, cast out the bones wrapped in the fur or feathers of the prey. By examining their cast pellets, a researcher can determine the species of prey taken by an owl.

The crop of the pigeon and dove produces a "milky" food substance from cells that line it. This rich source of protein and fat is fed by both male and female to the squabs. The secretion begins during incubation and continues until 16 days after hatching. As far as is known, the pigeon family is the only one that feeds the young from body secretions, as do mammals from the mammary glands.

The stomach consists of two regions. The first, called the proventriculus, has special cells producing enzymes that break down proteins in the food. The size of the proventriculus obviously depends on the food habits of the bird and is longer in fish- and meat-eating birds than in other species.

The second section of the stomach, called the gizzard, might be considered a modification in birds that partially substitutes for teeth on their jaws. In all species of birds the gizzard serves as a further storage area; however, the gizzard is mainly a grinding organ. It is surrounded by powerful, well developed muscles outlined with a thick, heavy, keratinized membrane that protects its walls. In seed-eating birds such as parrots it forms a powerful grinding organ. The mechanics of grinding, however, is dependent on the bird's swallowing small grains of sand; or pebbles, in larger birds such as chickens. Seed-eaters swallow their teeth, so to speak. The sand or small pebbles are mixed with the enzyme-impregnated seed and the powerful contractions

The proventriculus, or first region of the bird stomach, is larger in meat-eaters, such as this orange-fronted falcon shown here with its parrot prey.

of the muscles grind the whole mass and break it down for further digestion in the intestine. Such particles are highly important; if I did not put clean sand in one of Sam Redsocks' food dishes, he would quickly die, unable to digest his hard sunflower seeds. Birds of prey, however, do not have well developed gizzards, and that organ is a mere pocket where the flesh of the prey is stored for further digestion by the enzymes.

The Important Intestines

As in all vertebrates, the bird intestine is long and convoluted and is the main organ for absorbing of the food into the bloodstream. The liver is large in proportion to the size of a bird's body. Like the pancreas, it empties its ducts into the intestine. The pancreas is very important in birds because it produces the enzyme that breaks down starch. Birds, unlike mammals, do not have starch-reducing enzymes in their saliva. The liver is an important glandular organ that secretes bile, and as we shall presently see, chemically ties the circulatory and digestive systems together.

The main difference between the intestine of birds and that of other vertebrates is that the long bird intestine is much better "tied down," so to speak. The adaptation of birds for flight requires that its organs do not move about inside the body.

Birds that take in a lot of cellulose—seed-eaters and woodpeckers, for instance—have two well developed pouches called caeca that open into the last part of the intestine, the small intestine. Caeca are involved in water absorption but are also the main site for the breakdown of cellulose by specialized bacteria. Caeca are not present in fish- and meat-eating groups and are considerably reduced in hummingbirds and swifts. They are absent in parrots also, despite the fact that the parrot group consists mainly of seed-eaters. They use their powerful beaks, however, to husk the seed of its hard outer shell. Pet books on parrots say little about the rather subtle behavior of these remarkable birds, but conures, cockatiels, and larger parrots are known to shred wood with their beaks. I keep a fresh limb in my conure's cage, which gives him an opportunity to chew and prevents an overgrown beak.

The rectum is the last segment of the digestive tract and functions largely to absorb water. The rectum empties its wastes into the final chamber, called the cloaca. The sexual organs too empty into the cloaca, and since it is also the organ of copulation it will be discussed in the chapter on reproduction and mating.

7/FOOD BECOMES ENERGY

Dying in the nest is a common occurrence among birds. It is caused by numerous factors but by far the most common is starvation. Birds have an extremely high rate of metabolism and cannot survive long if their food supply is limited.

A classic of ornithological literature is Jean M. Linsdale's *The Natural History of Magpies.* In that monograph she points out that food plays the most important role in the survival of nestlings. Magpies collect their food within a half-mile radius of the nesting site. They may have a large family to support, such as five to seven nestlings. If the supply is not equal to the demand, the last-born nestlings die of starvation or are killed by their stronger siblings.

Magpies are mainly ground-feeders and a late snow may cover up their food supply. A. A. Saunders noted that in Montana on the day following a late (June 8) snow, two out of four nestlings were dead. He concluded that late snowstorms cover up the food supply and produce high mortality.

The beak is the food collector and the digestive tract the processing system which converts food into physical energy and heat. It accomplishes the complex job by a process called metabolism, which depends on the circulatory as well as the digestive system.

Simply put, metabolism is the sum of all processes concerned in the building up of protoplasm. It includes the chemical changes in living cells by which energy is provided for muscular work, respiration, digestion, nerve conduction, and other vital processes, and by which organic material is changed to basic biochemicals for repairing worn-out body tissues.

After being processed by the proventriculus, which produces protein-digesting enzymes, and the gizzard, which grinds it up, food passes to the intestine, where the chemicals from the processed food are absorbed into the bloodstream. The starch-processing enzymes are added by the pancreas. The blood passes through the liver in thousands of small vessels, and during its passage picks up bile from the organ; this makes important changes in bloodstream chemicals. One such change is the conversion of sugars to glycogen, an energy-releasing source for muscles, which is stored for future use. Bile, which is acid in birds, also forms urea, an end-product of protein metabolism. The liver is larger in fish-eating and insectivorous birds than in seed-eaters.

The Circulatory System

The heart of a bird is highly efficient. It is similar to that of mammals but differs in having a heavily muscled left ventricle with a small, star-shaped inner space, or lumen, and a much less muscled right ventricle with a large, cross-shaped lumen. The left ventricle is three times the size of the right one, and composes almost the entire bottom portion of the heart. It pumps blood throughout the whole body. The right ventricle is developed upward rather than being side by side with the left ventricle, as in mammals. It pumps blood to the lungs alone.

Blood enters the right atrium (the lumen of the right auricle) from the veins leaving all the parts of the body and flows through a valve to the right ventricle, where muscle contractions force it to enter the pulmonary artery leading to the lungs. In the lungs the carbon dioxide is released from the blood and oxygen picked up by the red blood cells. Pulmonary veins take the blood to the atrium of the left auricle, hence to the left ventricle at the bottom. Here the thick powerful muscles contract to pump it throughout the body by way of the large aortic arch and arteries. A minute network of capillaries, so small that they pass only one blood cell at a time, connects the arteries back to the veins. Blood plasma and white corpuscles filter through the thin capillary walls and into space between the body cells feeding them. The remainder of this food-carrying capillary fluid flows back into the veins by way of the tiny ends of the venous capillaries, completing the blood-oxygen-food cycle throughout the body.

The aortic arch continues as the large dorsal aorta, which passes between the lungs and into the abdominal cavity. Branches carry the blood to the stomach and intestine, where the food is absorbed; and to the liver, pancreas, and spleen, where they contribute their enzymes and bile to the fluid mixture.

Certain tissue fluid that fills the space between the cell walls is called lymph. It is absorbed by special fine vessels called lymphatic capillaries. These increase in size and become large vessels, making a complete lymphatic system. The largest such ducts, which occur as a pair, are called thoracic ducts. They arise in the lower body and empty into the jugular vein in the neck. Absorbed fat from the intestine is processed by means of special lymph channels called lacteals and carried to the veins by way of these thoracic ducts.

The weight of a bird's heart is in inverse proportion to the body weight of the species. The heart of a hummingbird, for instance, is 2.4 per cent of its body weight, whereas that of a large bird of prey, such as the broad-winged hawk, is only .57 per cent of its body weight (see Table 3). The weight of the heart is dependent on numerous factors but is most certainly related to the activity of the bird. The easygoing, soaring broad-winged hawk has a much less intense metabolism than the hyperactive hummingbird with its high wingbeat. A fair generalization is that ground-feeding birds and soaring or gliding birds have smaller hearts than fast-flying birds such as falcons and sandpipers.

TABLE 3

Weight of birds' hearts as a percentage of total body weight

SPECIES	% OF BODY WEIGHT
bobwhite	.39
brown pelican	.81
black vulture	.90
song sparrow	1.18
crow	1.20
killdeer	1.35
hummingbird	2.40

Salt and Water

There is much we do not understand about metabolism. Students of bird life observe many peculiar behavior patterns in birds. William R. Dawson and his colleague at the University of Michigan in Ann Arbor studied the peculiar salt use of red crossbills. Certain desert and salt-marsh-dwelling birds such as the Savannah sparrow, which inhabit environments where only salt water is available, have evolved a system for making use of such water. In other words, they are highly salt-tolerant. Red crossbills are often observed feeding on pure salt. This species is highly mobile and wanders across the entire northern states. Dr. Dawson trapped his test birds in the Huron Mountains of northern Michigan by attracting them to salt blocks. He fed them various amounts of salt solution and found that the higher the salt concentration, the greater the fluid intake in relation to body weight. In spite of the high salt content of the water, they remained in good condition until the intake reached .255 milliliters a day, at which point they began to respond erratically. A serious loss of weight began at .300 milliliters a day.

Strangely enough, European finches such as the house finch have the ability to tolerate salt water. Studies by Dr. Knut Schmidt-Nielsen and Dr. Ragner Fänge have shown that sea birds are able to excrete, by way of the kidney, only about one-half of the salt found in sea water. They discovered a salt gland just above the eye that can secrete fluid with a higher salt concentration than sea water has. This salt gland is found in many salt-water species as widely evolved as penguins and pelicans.

Apparently desert-inhabiting birds such as mourning doves and California quail have evolved highly efficient salt-handling kidneys that allow them to drink at salty water holes. Birds lose very little water by excretion because most of the water they drink is reabsorbed by the kidneys, leaving the semisolid uric acid salts to be excreted in the feces. Excess salt in mammals is excreted by the kidneys and sweat glands.

Dr. Oscar Johnson of Moorhead State College, Minnesota, studied the size and shape of 181 species of birds from 20 orders and found that the smaller birds have larger kidneys in relationship to body weight. This probably reflects the fact that smaller birds in most cases metabolize more actively than larger birds.

Since the crossbill is primarily a forest bird, and not a desert species, it is interesting to speculate that crossbills may have originally evolved in a hot, dry climate but have retained their highly efficient salt-processing kidney. Crossbills and other finches have been observed by birders feeding on road salt and salt from ice cream freezers. Salt-feeding seems to be confined to seed-eaters, so we may presume that the behavior evolved in desert ground-feeding species as a mechanism for maintaining their ionic salt-water balance in the body.

Exchanging Gases

Birds are unique among all vertebrates in that oxygen-carbon dioxide exchange takes place during both inspiration and expiration. This is due to a complex network of capillaries that allow for continuous circulation of air throughout the lung cavities. These connect to the right and left bronchi, or paired breathing tubes going to the lungs. The trachea, or main breathing tube, is located below the esophagus and is supported by a series of tough, well developed cartilaginous rings. The entrance to the trachea is surrounded by the cartilage of the larynx. The opening is called the glottis.

The other channel for inhaling is the nasal cavity. In birds the openings are at the base of the bill and are called nares. They are simple openings but are modified into a horny tube in some birds such as the albatrosses—thus the Latin order name, Tubinares, or tube-nosed swimmers (now often called by the order name Procellariiformes). Anhingas (water turkeys), cormorants, gannets, and such underwater birds have closed nostrils and presumably breathe only through the mouth and glottis.

Birds are also singular in that, like insects, they have air sacs. There are nine pairs located throughout the body. These are closely tied into the parabronchial network. Pouches lead from the air sacs and extend into the bones, which, as we know, are hollow (pneumatized). The amount of air-sac space and pneumatized bone differs with species but regardless of the size the system contributes to buoyancy.

Most bird anatomists consider the air sacs to be an expansion of the bronchial walls. The sacs are made up of very thin membranes transversed by elastic fibers. Since they have no

CATHY CALLAHAN

Birds have a high body temperature. This anhinga, or water turkey, is spreading its wings in the sun to dry itself out. This is important, for if it becomes waterlogged and thus chilled it will soon die.

muscles or blood vessels they cannot contract or absorb oxygen; therefore they are not directly involved in respiratory gas exchange. In spite of being without capillaries, they receive nourishment by contact with lymphatic fluid.

Thermoregulation: the Control of Temperature

The rate of metabolic processes in living bodies is determined by temperature. Organisms that keep a basically constant body temperature, as birds and mammals do, are called endothermic. Endothermy may be defined as the ability to maintain an essentially constant body temperature regardless of the surrounding temperature. Lower vertebrates, such as fish, amphibians, and reptiles, which basically have temperatures that vary according to environmental warmth or coolness, are called ectothermic; likewise insects and other invertebrates. (Some ectotherms such as lizards can control their temperature somewhat by their frequency and method of basking in the sun or seeking shade.) Endothermy gives an animal flexibility and is a prerequisite for superior mental ability.

Birds have a very high metabolic rate and corresponding high temperature. The internal temperature of a bird, depending on species (see Table 4), ranges from 40° to 45° C (104° to 113° F). The mammalian range is from 36° to 38° C (97° to 100° F).

Generally, large birds have a lower metabolic rate than small birds. A large bird such as the golden eagle may expend only 34 kilocalories per kilogram of body weight (kilo = 1000) whereas a pigeon expends 105 kilocalories per kilogram and a small hummingbird 1601 kilocalories. Such calculations show that the rate of metabolism is inversely proportional to the weight of the bird. Large birds produce less heat per unit of weight than small birds.

TABLE 4

Internal temperature of certain bird species
(FROM JEAN DORSET)

SPECIES	MAXIMUM BODY TEMPERATURE (IN DEGREES CELSIUS)
ostrich	40.0
white swan	41.0
quail	42.0
mourning dove	42.7
tawny owl	41.0
ruby-throated hummingbird	38.9
cardinal	42.5*
house sparrow	43.5

* Among vertebrates there is a daily temperature rhythm. The temperature of the cardinal, for instance, may range from as high as 42.5° C during daylight activity to as low as 38.5° C at night. Nocturnal birds have a reversed rhythm.

The basic rate of metabolism changes, of course, with activity and season. Thus an active bird, or a bird during migration, requires much more food in order to keep its metabolic rate and temperature stable. The metabolic rate is also increased during molting and breeding season. It has been demonstrated that while the remages (tail feathers) of a chaffinch are growing, its basic metabolic rate increases about 25 per cent. An active bird, whether molting or migrating, has a better chance of survival if it is well fed. A well fed sparrow, for instance, can withstand a drop to 21° C (70° F)—half its normal body temperature—but a starving sparrow dies if its internal body temperature falls below 32° C (90° F). That is why the fledgling magpie did not survive when snow covered the ground food supply.

Birds have no sweat glands, so they regulate loss of heat

Birds and humans use the same principle to keep heat in—or in some cases, out. This barred owl is puffing out its feathers in cold weather; the air trapped between the layers of feathers makes an effective insulator. The thatched roof of this Japanese farmhouse operates in the same way; the house is far warmer in winter and cooler in summer than it would be with single-layer roofing.

by evaporation from the respiratory system through their mouths. Feathers are one of nature's most efficient insulating materials, and birds, like mammals with hair, shed or grow feathers according to the external temperature. A Carolina chickadee in the winter may be covered with as many as 1704 feathers but by June have only 1104.

Birds also control heat loss by certain behavioral patterns. Most nature lovers have noticed that birds at the winter feeder tend to look all "puffed out." Birds control the body feathers with dermal muscles. Ruffed feathers increase the thickness of trapped layers of air near the body, thus exposing less surface to cold air and making a more effective barrier to the outward radiation of body heat. They also make the bird look somewhat spherical. Many species roost together during cold weather and thus pool their body heat. I will never forget one snowy day in the Ozarks when I was reaching into holes in dead trees in order to check for screech owls. From one hole I pulled out six tufted titmice all huddled together for warmth.

Feet and Legs as Thermal Controls

Birds lose considerable heat through their feet and legs, which are not covered by feathers. One would imagine that ducks, for instance, paddling around in an icy pond would, so to speak, freeze their toes! Such is not the case, for birds' legs have greatly reduced blood circulation. The ends of the toes are, in fact, almost the same temperature as the air, or water in the case of ducks; this prevents heat loss. Ducks have heat-exchangers in their legs and toes in the form of an arteriovenous network. The fine network lies between arteries that carry warm blood to the toes and veins that carry cool venous blood back to the leg from the toes. During hot weather the feet radiate excess heat back to the environment. This is accomplished by adjustment of the arteriovenous network to allow an increased flow of warm blood toward the feet (which are at body temperature) and thus dissipate the heat.

Regardless of what physical or behavioral means of temperature control is used, thermoregulation in birds is highly efficient and body warmth is maintained at the temperature most suitable for the metabolic processes of the species.

8/ THE VERTEBRATE NERVOUS SYSTEMS

In order to get the proper perspective on birds as the complex, flexible creatures they are, it is necessary to look at the nervous systems of vertebrates in general, even going as high as human beings. The vertebrate nerve arrangement is generally divided into two parts, according to what the nerves do. The brain and spinal cord make up the central nervous system, which is the main control. Nerves that exit from the brain and spinal cord and go to other parts of the body are called collectively the peripheral nervous system.

The nervous system as a whole controls and integrates all the activity of the body. Through sensory organs of sight, sound, touch, smell, and taste, it receives and interprets stimuli from the environment. Birds depend strongly on sight and so, as might be expected, have well developed optic lobes in their brains.

A major difference between the brains of mammals and reptiles and those of birds is that the large optic lobes of birds are located at the sides of the brain instead of on the roof of the midbrain. This modification probably represents the need for a highly developed visual system in flying animals. The side position allows for expansion of the lobes and efficient placement of the eyes for peripheral vision. Seeing at acute angles from all sides of the eyes is very important, for birds in their swift flight have no time to turn their heads constantly to avoid hitting things above or below them.

The cerebrum, or upper forebrain, of a vertebrate consists of two distinct types of nervous tissue: white matter, which is made up mainly of nerve fibers (axons or dendrites, which are thin growths from nerve cells, or neurons), and gray matter, made mostly of neurons plus some nerve fibers. Large amounts of gray matter are generally associated with the superior thinking powers of the more highly developed vertebrates, found at the top of the evolutionary tree of life. Birds have more gray matter than reptiles and are placed by behaviorists between reptiles and mammals in their ability to learn.

We still do not know how organisms "learn," as far as the brain is concerned. Forty behaviorists, physiologists, and psychologists could write 40 volumes on the subject and all would probably disagree, to at least some extent. How much any group of organisms can learn depends on the amount of gray matter in the brain. The great paleontologist-geologist Pierre Teilhard de Chardin, in his book called *Man's Place in Nature*, calls the gradual evolutionary increase in complexity, as we move to higher branches of the tree of life, the process of cephalization. Cephalization, which is measured by the amount of gray matter and size of the cerebral hemisphere in relation to body weight, reaches a climax in human beings. Teilhard believed that the higher the organism the greater is a factor he called "psychic temperature" (consciousness of self and of the surrounding environment). Of course no one can really define the word "psychic." It is a word like "instinct": used to describe that which we still do not understand. The difficulty with studying the brain, and in particular the cerebral hemisphere, is that one is using the instrument to study the instrument, which may well be impossible!

If the optic lobe of birds is highly developed, the olfactory lobe is, on the other hand, only slightly developed. Most birds,

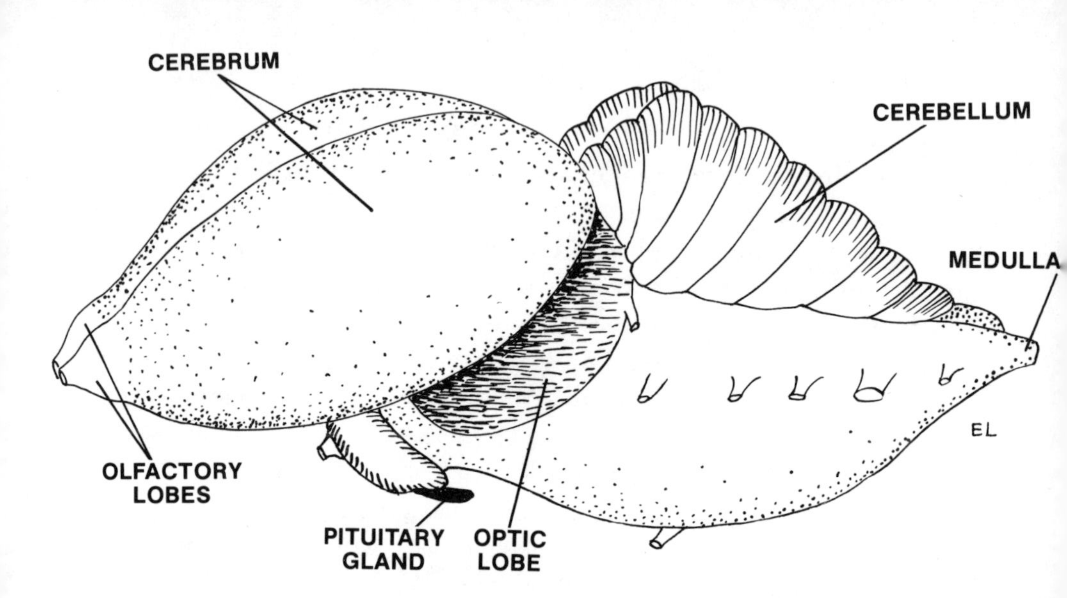

The brain of a chicken. The optic lobes in a bird of prey are relatively larger. The eight stubs are the cut-off beginnings of various cranial nerves.

as nearly as we can determine, do not depend much on their sense of smell. Olfaction is of course highly developed in most mammals.

The other main parts of the brain, other than the cerebrum, are the cerebellum, which is located dorsally at the base of the head, and the medulla, which is located beneath the other parts. The cerebellum is the portion of the brain that coordinates the voluntary muscles and maintains equilibrium. The medulla, also called the medulla oblongata, has an oblong shape and connects the spinal cord to the brain. It contains most of the major fiber tracts and nerve cells between the brain and spinal cord and is responsible for controlling the involuntary vital functions such as breathing, blood flow, and blood pressure.

The Peripheral Nervous System

The peripheral nervous system consists of two main groups: the cranial nerves connecting the brain and other parts; and the spinal nerves, plus their motor and sensory endings, that connect the spinal nerve cord and the whole body. Unlike mammals, which have 12 cranial nerves from the brain, birds have only 11. Most of the 11 cranial nerves go directly to the brain but one, called the vagus, or wandering nerve, connect the heart, lungs, and stomach to the brain.

The nerves that feed the smooth (involuntary) muscle, heart muscles, glands, skin, blood vessels, and visceral tissue are composed of motor fibers and lumped together under the title autonomic nervous system. They supply those systems which operate regardless of whether or not we are "thinking" about them. In other words, they are mostly automatic.

Biofeedback

The so-called "facts" of one decade of science are sometimes the nonfacts of the next decade. Science moves so swiftly that each day brings new discoveries. For years researchers have marveled over certain persons, Indian fakirs for instance, who can voluntarily control certain body functions, as by lowering their heart rate. The phenomenon seems to indicate that the autonomic nervous system is not completely automatic. Thus the qualifying "mostly" above.

The vertebrate brain puts out specialized electrical currents called brain waves. Beta waves, from 13 to 25 cycles per second (hertz), are associated with normal, focused, rational thoughts in humans; alpha waves, from 8 to 13 hertz, with nonfocused daydreaming; and theta waves, from 4 to 8 hertz, with a twilight state, or condition of reverie usually experienced just before sleep and associated with high creativity. In birds, theta waves signal not only the sleep state but also what is known as torpor, a state of inactivity and sluggishness. Scientists have been able to detect these waves and, by converting them to sound, feed them to human subjects' ears. Subjects can learn to control their own waves by slowing their breathing and relaxing. As the waves drop, the subject reaches the relaxed 4-to-8-hertz theta rate and becomes, so to speak, "high" without drugs.

The whole process of biofeedback is tied to what are called the sympathetic nerves, which are part of the autonomic system.

The Autonomic Arrangement

The autonomic nervous system is divided according to its physiological functions into the parasympathetic and sympathetic systems. The sympathetic system speeds up the bodily functions. It increases the blood flow to all systems, especially the brain, so that hearing and vision become more acute. It is, for instance, the precursor state for a falcon about to dive on a prey. The parasympathetic system slows biological activity. It takes over during sleep, and the appendages, hands and feet, tend to warm up. We might say that it can create a sort of twilight state.

The parasympathetic and sympathetic systems are tied directly to and control the metabolic rate of living organisms. Scientists believe that biofeedback works through hormones that circulate through the body.

The emotional part of the brain, called the limbic system, triggers the master control center of the brain, the hypothalamus. This in turn triggers hormonal production by the pituitary gland, which is one of the master control glands in all vertebrates. It will be discussed in the next chapter. All of these complex physiological functions depend on thermoregulation, which is particularly critical in birds because of their high body temperature.

Some species can survive a severe drop in their internal temperature. Anna's hummingbirds live in our western desert and go into torpor when the night temperature falls. The little birds find a hiding place and their body temperature drops as low as 19° C (66° F). There are no detectable reflexes, so the hormonal control of the parasympathetic system slows the entire body almost to a standstill. Whether or not the state of torpor can be classified as an extreme form of biofeedback phenomenon, as in humans, is debatable but certainly both result in an extremely low metabolic rate.

Seeing the World

The eyes of birds are remarkable structures and have been studied extensively. Their acuity in seeing distant objects with great clarity depends on resolving power—that is, the ability of an eye (or for that matter, a camera) to show distinctly the very fine details of an object at a distance. The farther away the object is, the greater is the resolving power needed to show it up distinctly against its background and objects next to it. In human beings, the limits of resolving power can be demonstrated simply with an eye chart. The smaller and more crowded the letters of such a chart, the more they appear to merge and become, in effect, linear black blurs; letters as such can no longer be seen.

The remarkable avian resolving power comes from two factors. One is the size of the eyeball, lens, and retina—substantial size allowing a large image to be formed on the retinal screen. The second factor is the great number of light-sensitive cone cells in the retina and the high ratio of optic-nerve fibers going from them to the brain. The retina is a tissue at the back of the eye that acts as a receiving screen for the image and that translates this image into nerve impulses. It contains two types of cells: rods and cones. The rod-shaped cells are very responsive to faint light but not to color. The cone-shaped cells respond to stronger light and to color. Cones are found all over the retinal surface and are extremely concentrated in the fovea, which is a depression in the retina of each eye.

The Marvelous Screen: the Retina

Nocturnal birds, as would be expected, have more rods than cones, since they have evolved for activity in very dim light. The cones of daytime birds contain peculiar oily globules colored brilliant red, yellow, orange, and green. The globules are colorless in nocturnal birds.

The dense concentration of cones contributes greatly to the resolving power of the bird's eye. Some broad-winged (buteo) hawks have up to one million cones per square millimeter of retinal surface. Up to 120,000 cones per square millimeter have been counted in the retina of the wagtail's eye. The human retina contains a mere 10,000 per square millimeter.

The optic nerve which leads from the brain to the retina differs from other cranial nerves in that it develops directly from the embryonic brain. It may be considered closer to being a fiber tract (a bundle of nerve fibers that have a common origin, ending, and function) of the brain than a nerve cord.

The eyeball of a bird, which includes both the retina and lens system, may equal or exceed in size the entire brain of the bird, being larger in proportion to body size than other vertebrate eyes. Birds are, so to speak, "all eyes."

The Fovea and Pecten

Specialized areas of the retinal cone cells that increase in density to over a million cone cells per square millimeter, and where there are no rods, are called foveae. There is a depression and thinning of the retina at these spots. These areas make vision of extremely great resolution possible. The shape of the foveae differs among various groups of birds. In ground-feeders, such as quails, it occurs as a rounded central area. In water birds that inhabit open areas it is centrally located but much more elongated. Birds of prey and other birds such as swifts, swallows,

The eyes of raptors, or birds of prey, are particularly sensitive, acute information-gatherers; the visual acuity of some hawks and eagles is said to be at least four times as great as that of humans. This is a red-tailed hawk (black phase).

The excellent eyes of birds sometimes have extra protection. Some raptors, for instance, have a bony ridge above each eye. This road-runner has long, wiry modified feathers near its eyes; these help protect the organs against thorny cacti and bushes.

and flycatchers that catch prey on the wing have both the central fovea and an extra one called a temporal fovea placed laterally toward the side of the eye. The light rays that pass directly into the eye fall mainly on the central fovea of most birds, whereas in the binocular vision of the prey-catching types the light falls on the side fovea.

The pecten is a membrane located at the back of the eye. It contains numerous blood vessels and projects into the eyeball from the optic nerve. There are almost as many guesses as to what is its function as there are ornithologists. Most researchers feel that it functions mainly to nourish the retina with blood. It is highly unlikely that this is the only function of the pecten. Two ideas are that it compensates for changes in pressure in the eye brought about by movement, and that it actually enters into vision and increases the sensitivity of eye movement.

Since it is a thin convoluted membrane, I consider it to be a membrane *electret,* or polarized insulating body, that (along with the bony ring that surrounds the eye) is a part of a dielectric antenna system. (A dielectric is a nonmetallic, insulating substance in which a steady electric field can be set up.) In my opinion, the pecten orients the bird to the fields that surround the earth—and here I refer specifically to the electric fields, not the magnetic fields. I consider such electric fields to form "electrical" paths along which birds migrate by using an "eye dielectric antenna" system—the pecten and its connections to the brain. Since no one has really studied the electrical properties of the pecten in any detail and since no one has ever measured any electrical highways around the earth, I realize that most scientists will consider my theory to be far-fetched.

Hearing

The ear of a bird is located on the side of the head under a group of feathers called the auricular patch. Birds have no

A young black vulture, in which the ear opening, usually covered in birds, can be seen to the left of the eye. There is a large nostril in this species, which has a keen olfactory sense.

external ear flap, or outer ear, as mammals do; the auditory canal simply emerges at the side of the head and stops. An eardrum is found near the inner end of the canal. The ear of the bird is a very important sensory organ, as song is one of the most important means of communication. Owls hunt at night and are successful in catching their food in complete darkness, thanks to specially adapted ears. These are different on either side of the head, each having its own size and shape, and one being lower on the head than the other. Such variations enable the bird to pinpoint the mouse or other prey no matter how extremely faint the sound of its movements, and are aided by a special method of processing the signals as they pass through its nervous system.

Birds have a voice box called a syrinx. In mammals the voice box is called the larynx and is located at the top of the windpipe, or trachea. The bird syrinx, on the other hand, is at the lower end, where the trachea branches into the two bronchial tubes leading to the lungs.

Smell

Birds have poor olfaction powers. The olfactory bulbs of the brain lie in front of the cerebral hemisphere and are much smaller proportionately than in mammals.

The nasal cavity is separated by a septum, or wall, and each cavity is divided into three chambers by folds. The middle chamber opens into the mouth by an inner nostril. The nasal septum is covered with a thin membrane called the epithelium, which has a growth of many cilia—minute hairlike projections that act as filters. The septum not only filters the air but also warms and humidifies it.

In spite of their complex olfactory apparatus, birds seem to be almost without a sense of smell. This sense has been studied in several species with contradictory results. However, condors, black vultures, and turkey vultures have been proved to use the sense of smell to help find their carrion food, though they too depend mainly on vision for this task.

9/ LIGHT, HORMONES, AND FLEDGLINGS: REPRODUCTION

Animals that reproduce by laying eggs are said to be oviparous. The word derives from *ovum*, Greek for "egg," and *parere*, Latin for "to give birth." Birds, as might be expected of creatures descended from reptiles, are oviparous. Mammals reproduce by viviparity, a word from the Greek *vivi*, "alive," meaning "live-born." Oviparity is not a very safe method of reproduction when compared to viviparity, in which the fetus is protected by being carried inside the body. Eggs are exposed to predation and external environmental factors such as heat and cold; also, eggs usually require a nest, which limits the mother bird's activity. The sedentary nest period exposes her to danger. The advantage of oviparity is in economy of weight for flight.

The Female Reproductive System

Strangely enough, the female bird's reproductive system is not paired and symmetrical, as in mammals. The right ovary does not function and is rudimentary, a characteristic which further lightens the bird. However, in some falcons and hawks there are two ovaries. In our drawing the ovary at left is shown in the breeding condition; the right ovary is in the much reduced

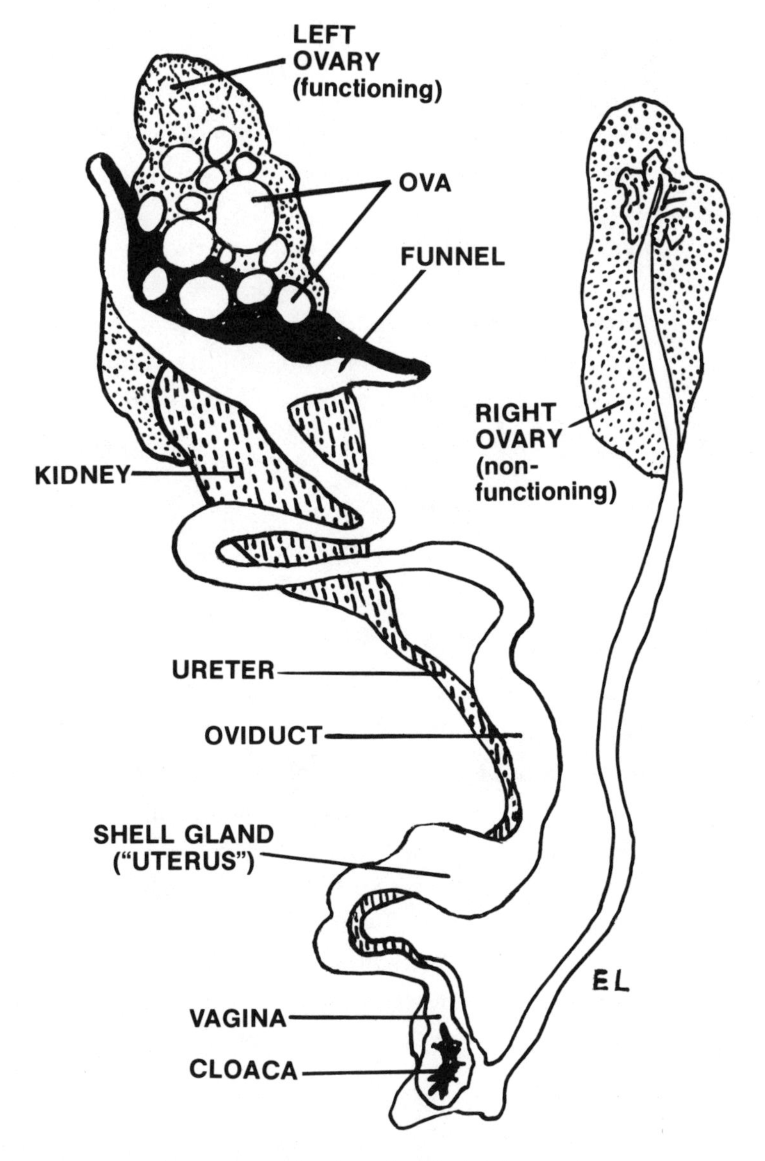

The reproductive system of a female bird. The mature eggs, or ova, drop into the funnel (infundibulum) and are brought to their final form along the route to the cloaca.

nonbreeding condition (eclipse). Ovaries are located in the abdominal cavity in front of the kidneys.

During breeding season the ovary enlarges and the immature eggs and follicles (sacs that enclose them) resemble a bunch of grapes. Only a few ova (mature eggs) are released from the numerous follicles during ovulation. Once freed, the eggs are fertilized by the male sperm and pass into the wide funnel-shaped infundibulum and down the oviduct to the uterus. An ovum at first carries the yolk alone. It picks up albumen (egg white), which is an additional food source, in the middle area of the oviduct where the white is produced. The small membrane and shell are formed near the exit, and the completed egg is moved through the vagina and laid from the cloaca, which is a combined anus and egg-depositor. The passage of the egg through the oviduct takes about one day in most species.

The Male Reproductive System

Unlike those of mammals, in which the testes are located in an external scrotum, the bird testes are inside the main body cavity, in front of the kidneys. This has the advantage of streamlining the male body. To compensate for the increased heat, which might inhibit spermatogenesis (the formation of sperm) the testes move backward in the abdominal cavity during the breeding season and are then positioned between abdominal air sacs that serve as a natural air-cooling system for them. During the breeding period both testes of a male may enlarge from hundreds to thousands of times, depending on the species. Light, working through the endocrine system, is the main control for such changes.

The sperm from an enlarged testis (on the left in our picture) move down into the epididymis and then through the deferent duct (sperm duct) to the cloaca. The ureters, which

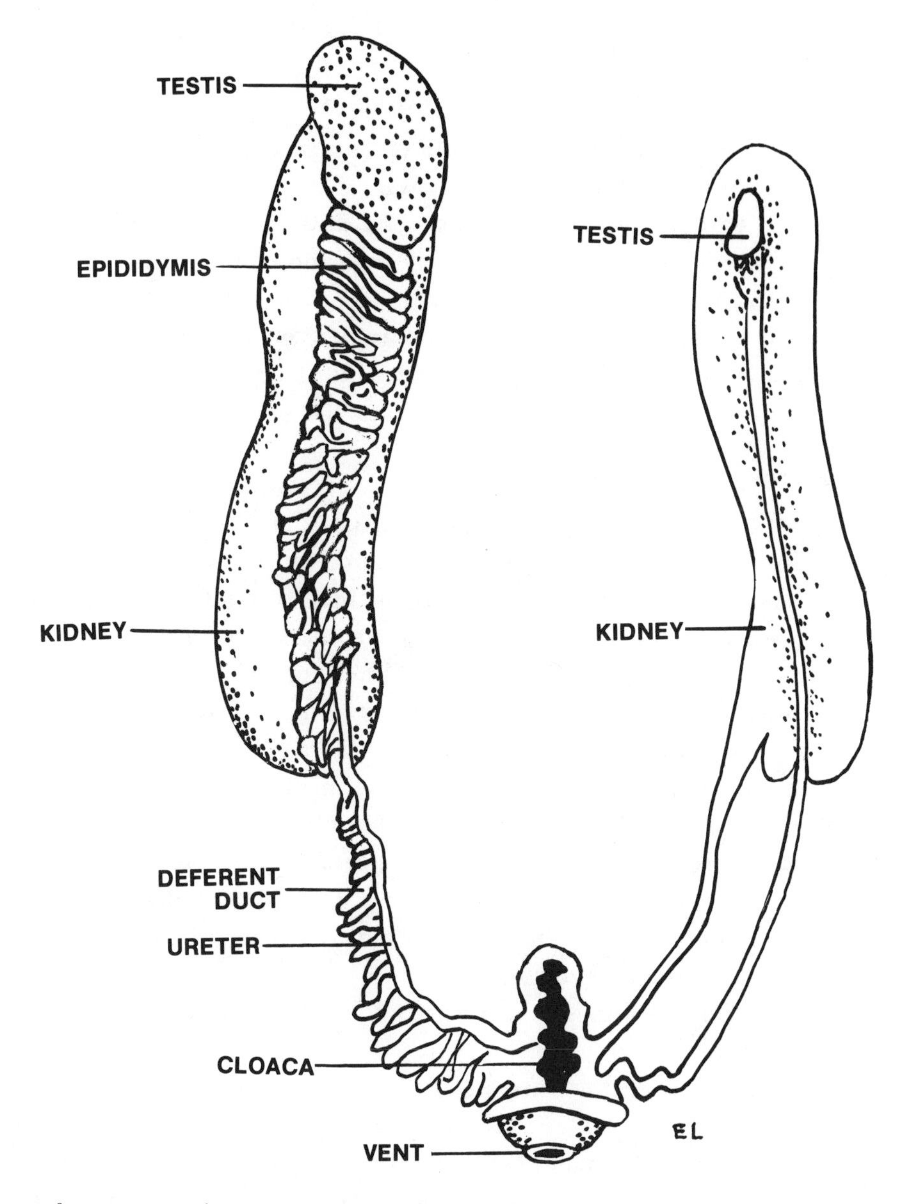

The reproductive system of a male bird. On the left the testis and associated organs are shown enlarged and well developed at breeding season; at right the smaller size during nonbreeding seasons is shown. (It should be noted, however, that in some birds the two testes normally differ in size and shape, regardless of season.)

carry excretions from the kidneys, run parallel to the deferent duct and all empty into the cloaca.

Display behavior during courtship leads to mating, or copulation—the sexual union of male and female that transfers sperm from the cloaca of the male to that of the female. Unlike mammals, male birds have no penis. The male bird balances on the female's back and then presses his cloaca against hers.

Sexual Maturity

Various bird species mature sexually at considerably different ages. The bald eagles, for instance (our national emblem), reach reproductive ability at about three years. Their beautiful plumage, with white head and tail feathers, develops at the end of their third annual molt. Actual mating and breeding do not start before the age of four, and may be as late as age six.

Birds of smaller size—woodpeckers, sparrows, doves, for instance—start reproducing the first spring after they have left the nest in which they developed. Larger birds (such as owls, sea gulls, geese) begin their reproductive activities the second year after abandonment of the nest. Dr. J. D. Ligon, an ornithologist who studied the endangered red-cockaded woodpecker in Florida, found that the majority of them did not breed when one year old; this was true especially of the males. But one-year females often did breed at this age, after wandering until they had discovered an unmated male that held a territory. The fledgling woodpeckers stay with their parents until the following spring, and the males thus have little or no time to establish a territory or to find a tree sufficiently rotted inside to be excavated for a nest. A proper nest and a territory well supplied with food must be set up, however, before a pair bond can be formed with a free female, so there are complications in the reproductive life of this species.

Reproduction as a process moves along through a complicated set of behavioral patterns, each of which must be followed through for success. These patterns depend on the endocrine glands of the birds.

Chemical Triggers: the Endocrines

Sexual activity, nest-building, molting, song, and migration are all under control of various hormones from the endocrine system. The ovaries and testes, besides furnishing the eggs and sperm, are endocrine glands and control the secondary sexual characteristics of the bird. All sexual creatures exhibit sexual dimorphism—the existence of two distinct forms, male and female. The maleness and femaleness of an organism are said to be primary sexual characteristics.

In most cases the male and female bird can be distinguished by what biologists call secondary sexual characteristics. These are external traits which are easily visible to the observer, such as color, size, or morphological differences such as the comb of the rooster. The sex of many species is quite evident from color. The females of the blackbird family are dull brown while the males are usually black with iridescent purplish or blue tinges. The females of most raptors, or birds of prey, are approximately one-third larger than the male. Male warblers are more brightly colored than the female.

There is also a phenomenon called seasonal dimorphism. During breeding season many species put on a nuptial plumage. The male birds of paradise from New Guiana grow elaborate decorative feathers which they display to their female partner during courtship. Warblers and a few other species of passerine birds show seasonal dimorphism. The male and female of the black-headed sea gull lose the black head in the winter. Such a nonbreeding winter bird is said to be in eclipse plumage.

It is the hormones from the gonads that control these various secondary sexual characteristics. The word "hormone" derives

The preparation of nests, which is triggered by hormones, can result in a great variety of fostering places. These least bittern chicks are in a loose nest of reeds; the killdeer eggs below are in a nest composed only of stones. If you look closely you will see that whiter stones have been collected for the actual nest site. The mother is making the best use she can of solar energy, for the whiter stones near the eggs will reflect more heat rays at them. The mother can leave the nest during the day and the gravel incubator will keep them sufficiently warm.

from the Greek *horman,* meaning "to stir up or set in motion." Hormones are specific organic chemicals produced by living cells and transported by body fluid. They bring about specific effects on the activity of cells remote from their place of origin. The two main sexual hormones are testosterone, the male hormone, and estrogen, the female hormone.

Birds are excellent experimental subjects for the study of hormone action. By removing the gonads or injecting hormones, physiologists can distinguish the interaction of genetic differences from the characteristics controlled by hormones.

Dr. Eldon Greij studied the effects of sex hormones on the plumage of the blue-winged teal. It has long been known that the testes control the eclipse plumage of ducks. Researchers had been able to feminize mallard male plumage by administering estrogen but were not able to influence feather patterns of the female with testosterone injections. Dr. Greij used an estrogenic hormone called diethyl stilbestrol.

This suppressed feather patterns in both sexes. On bare spots of skin, six milligrams of the hormone induced plain feathers (nonmale) in immature males identical with those grown by females. The feathers were the same pattern as those grown in wild birds in June when the estrogen levels are high. The hormone inhibited the molt of the birds but did not affect the regeneration of new feathers. The conclusion of course is that diethyl stilbestrol not only inhibits feather loss during the nesting season, when that would be an advantage for a nesting teal, but also keeps the plumage in the dull brown-colored hue so necessary to make the ground-nesting teal inconspicuous to predators while incubating eggs on the nest.

Testosterone did not modify the regenerated feathers on the bare spots of males; they remained dull. Adult female plumage is thus always brown but the developing plumage of adult male teals seems to alternate unless the estrogen level is high, in which case it retains the brown female color.

Hormones are an essential factor in the reproductive survival of the species. They are, so to speak, the traffic officers of the reproductive body, starting or stopping the process at appropriate times and thus matching the reproductive activity of the bird and its surroundings.

10/ THE POWER OF THE WAVELENGTHS

In most quarters of the world, the so called ignorant natives are far more knowledgeable about nature, and its mysterious ways, than city people. The training of elephants is a case in point. American zookeepers have been led to believe that the very best, and in fact only way, to train elephants is to "put the fear of God" in them with a pointed elephant hook. My experience with birds is considerable and with elephants slight, but I am certain of one fact: an elephant that is trained with kindness and reward is a far safer animal than one trained to fear its keeper. A fear-trained elephant will turn and kill its handler eventually —that is why elephants in America kill more handlers than do the work elephants of Thailand and Burma. Work elephants are treated gently.

In the late 1940's I took a long hike around the world. In the northern hill forests of Thailand I observed something about elephants and candles that may well introduce a discussion of birds and their reactions to light: that not only are elephants unafraid of fire but it seems a possibility that their physiology may be directly responsive to candlelight. Forest officers tell how elephants sleep standing up like horses. Of course all circus people know this, but in the forests where work elephants are free to roam they have been observed in their native habitat.

A work elephant in the teak forest of northern Thailand. Elephants lie down only briefly unless they are sick or exhausted.

During that time of night when the forests and jungles are suddenly silent, elephants actually lie down in a deep sleep for half an hour or so.

It is a period in the forest night when the leopard no longer calls and insects cease their noisy serenade. A strange, eerie silence envelops the woodland and, although the time varies somewhat, it most often occurs just before dawn starts the birds to voicing their territorial songs. It is a time of deep sleep for birds and mammals, but certain species of moths are stimulated to fly by the eerie ultraviolet that is invisible to humans but luminates the sky at that time of night.

Elephants that "go down" at times other than this "hour of the wolf" are a worry to their masters and forest officers. It is an indication that they are sick or so fatigued that they may never rise again. The forest people then burn large temple candles around such an animal—one for every year of its life. The elephant is, so to speak, bathed in hydrocarbon fire. The Westerner is skeptical of such procedures, but the forest officers do not forbid the practice. The treatment gives the impression that it always works on fatigued elephants; the animal soon rises completely refreshed. These treatments are of course not controlled experiments; one must decide for oneself whether cause and effect are seen here.

The mountain people believe that a spirit of the body leaves during deep sleep and thus one should be awakened slowly so that the "dream spirit" will not be caught napping in another land—it might not be able to return in time to the body, and the spirit-beast of life would be lost forever.

In the forests of southeastern Asia the line between superstition and reality is very thin indeed. The hill tribes attribute the power of the temple candle to the Nat gods of the forest who drink the light into their own spirit-bodies and so cure the elephant in gratitude. I personally am convinced, however, that the elephants actually absorb beneficial energy from the candles.

Candles have long been used in religious ceremonies, and medieval people understand the calming powers inherent in a flaming candle. Many of the last 20 years of my research life has been spent studying the candle flame. The flickering light of a candle is attractive to not only night-flying moths but to mosquitoes and certain ants as well.

JAMES BROGDON

A candle's light attracts night moths powerfully, and certain other insects as well.

A wild captured falcon is called a passage bird because it is usually taken during migration. Medieval falconers hooded such birds, not for cruel motives, but for the exact opposite reason—to suppress the bird's natural fear of humans. They fed the bird through the hood and gradually gained its confidence by removing the hood in subdued candlelight. This light has a calming effect on them; most creatures show little fear of the candle flame, despite the myth that animals strongly fear fire. A raccoon will approach a candle flame out of curiosity.

The electromagnetic energy that goes into living things lies mostly in the visible and near-ultraviolet rays that arrive at the earth's surface through the atmosphere. All we need do is look at the blue sky to get a sense of the power of light. Wavelengths of visible light are only half the answer, however, for organic molecules, stimulated by visible light, emit narrow-band wavelengths of infrared light similar to low-intensity laser waves. Not only do candles emit such infrared electromagnetic energy but so also does one's very breath. In many different native languages the word for "breath" is the same as that for "spirit," meaning something which many people suppose leaves a human body at death. All living organisms absorb visible and near-ultraviolet waves and emit infrared waves. That is why a candle is in a way so like a living organism. In the infrared region of the spectrum the candle emits many of the same wavelengths that our own bodies do.

In vertebrates, mammals and birds alike, the organ that controls and is the regulatory center for light is the pituitary gland.

The Gland that Directs

We saw in the last chapter that the female hormone injected into male ducks caused the growth of female feathering, whereas the reverse does not occur. The male plumage is believed to be basically characteristic of the species and is called the neutral plumage. It expresses itself in the absence of the female hormone. This is especially evident in old females of certain species that develop male plumage as their ovaries deteriorate.

The pituitary gland is the control center for the physiological processes that occur during plumage growth and gonadal changes. Among ducks, experiments indicate that the secretions from the male pituitary stimulate cock feathering. The female pituitary stimulates the ovaries to secrete the hormone that suppresses male feathering. The pituitary gland is located in the brain at the base of the hypothalamus. The seasonal size change of the gonads is controlled by two hormones from the pituitary.

This gland receives information from the hypothalamus, which is connected by relatively direct channels to the internal and external nervous receptors. One connection to the hypothalamus is through the eye. Ornithologists have long known that much of the reproductive activity of birds is influenced by light that enters the eye. The amount of light, received from the environment over a period of time is the controlling stimulus. As nearly as can be determined, the mechanism for this—and the pathway—lies in a group of special sensory cells in the retina of the eye.

Light falling on the eye starts the neurosecretory cells of the hypothalamus to producing a certain hormone; this in turn stimulates the cells of the anterior pituitary. The pituitary then releases the special gonadal hormones that cause the gonads to grow bigger and become mature. The cycle is completed when the gonads secrete the sex hormones, male or female, that control the sexual activity and feathering of the bird. This whole complex process begins and ends with the amount of light that "bathes the bird." Physiologists say that the bird is under the influence of photo-periodic control, or to put it my way, controlled by the power of light.

The Sun's Clock

Phenology is the science that relates organisms' regular bodily and behavioral changes to climatic events. Alterations in birds' gonads, or sexual glands, as they come near their breeding season are phenological events, like the appearance of corn ears in late summer, or the fall of leaves from trees in the autumn. A primary factor giving the "traffic signals" for breeding and migration in birds is light. Scientists give the name "photoperiod" to the alternating times of light and darkness that affect organisms' growing and maturing periods. Photoperiodism is the ability of an animal or plant to respond to changes in the amount or duration of light. In nature the length of daylight each day of course varies from the winter to the summer solstice. Scientists can investigate photoperiodism in the laboratory by building special chambers for animals, in which temperature and light are controlled according to the requirements of a given experiment; day and night are imitated or distorted to order, and heat is increased or decreased.

In the earlier periods of biological thinking, ornithologists believed that birds' breeding cycles depended mainly on variations in the warmth of their habitat. This seemed logical, because many birds migrate to warmer climates as winter comes on. Within recent years, however, light has been shown to be much more important than temperature in determining breeding schedules.

One investigator, Dr. C. M. Weise, studied the glandular responses of three species—white-throated sparrows, fox sparrows, and slate-colored juncos—to variations in day length. These birds were kept in an experimental chamber for 18 months with a continuous nine-hour "day" and a 15-hour "night" to imitate the shortness of days in winter. The gonads did not become any bigger under this regime, and the birds' usual reaction of premigratory restlessness did not appear in the normal way. Such response was delayed for two months in the juncos and fox sparrows, and the white-headed sparrows showed none at all.

Then Dr. Weise switched the cycle of night-day around to 15 hours of light and nine of dark in each 24 hours. With this program, the males' testes grew in weight from a two- to five-milligram level (at which they did not function for breeding) to between 200 and 300 milligrams, which is quite adequate for breeding. The ovaries of the females increased from a three- to eight-milligram weight to a level of 30 to 50 milligrams; this is the usual weight for females as they begin ovulation, or the formation of eggs in the body.

Thus biologists know today that the amount of light available from month to month through the year has a clear-cut effect on reproduction in birds. Unfortunately, however, very little research has been done on the *type* of light reaching them; that is, what wavelengths are involved. Wavelength studies in

The part of the electromagnetic spectrum covering low-frequency radio (10 kilometers, top left) through broadcasting wavelengths, microwaves, and infrared to far ultraviolet at 10^3 angstroms (top right); the hertz (cycles per second) are shown. Visible light is a narrow band of wavelengths between infrared and ultraviolet.

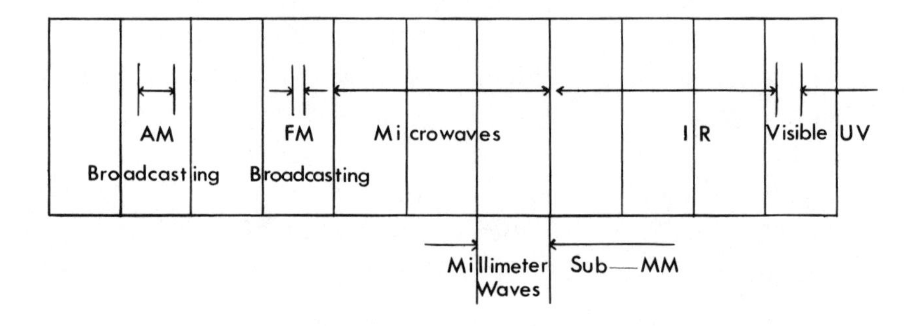

photoperiodism are few and far between. Visible light is composed of wavelengths from .4 to .7 micrometer in length. (A micrometer is 1/1000 of a millimeter.) Every color that we see in the spectrum that a prism can project has its own wavelength: for example, blue is .4 micrometer, red is .65 micrometer. The sun—and also various human-made devices—produces also two invisible kinds of light: ultraviolet and infrared. The sun, and also an ordinary light bulb, emit more infrared, in fact, than they do visible light. It is time for scientists to ask themselves much more often, what effects do ultraviolet and infrared have on birds' reproductive cycles?

An Experiment

In the space available here I can touch only briefly on the effects of light. But to get a clear-cut demonstration of what it can do, I ask my readers at this point to perform a simple experiment relating to light focused into the eye. It is an experiment that my friend Dr. John Ott uses to impress on his audiences the danger of constantly wearing sunglasses. The experiment emphasizes not only the power of light to stimulate but also the importance of knowing the exact wavelength of light.

Ask a friend to hold his or her stronger arm straight out in front at shoulder height. Take a sky-blue piece of construction paper some 8 x 14 inches in size and hold it about 18 inches in front of the subject's eyes for two minutes. At a "go" signal, ask you friend to resist as hard as he or she can while you try to push the arm down. Now repeat the same process but substitute a bright orange card for the sky-blue card. Again try to force the arm down after giving a "go" signal. In this experiment you can consider the amount of resistance to the downward force as an index of your friend's muscle strength. You must be certain to have very bright daylight or light from a fresh light bulb of at least 100 watts' strength falling on the blue and orange cards. The results of this test will surprise you and give you a feeling for the importance of wavelength and the power of light.

11 / BIRD TERRITORY AND BIRD "TALK"

We can think of territory and display behavior as being reproduction-control systems. Reproduction is a programmed, periodic happening. In most vertebrates—with the major exception of humans, who reproduce in all seasons—reproduction is intermittent. Among birds it is a spring phenomenon, and evolutionary selection has ensured that it takes place when environmental factors are such as to insure survival of the young. Temperatures are right and food is plentiful. As the Bible aptly puts it, "The voice of the turtle is heard in the land." The turtle in this case means the turtledove, the soft cooing of which is the harbinger of spring.

As we have seen, light and the endocrine system play a very important role in the seasonal cycle of bird life. We may think of the whole phenomenon as being a sort of closed cycle that repeats itself over and over year after year throughout the life of each individual bird. The details of each reproductive cycle vary, of course, from species to species but the general sequence of physiological events is the same.

We will not concern ourselves here with which scientist is correct about which cells in the bird's brain "collect" and act on the light; we will instead assume that the eye, pineal gland,

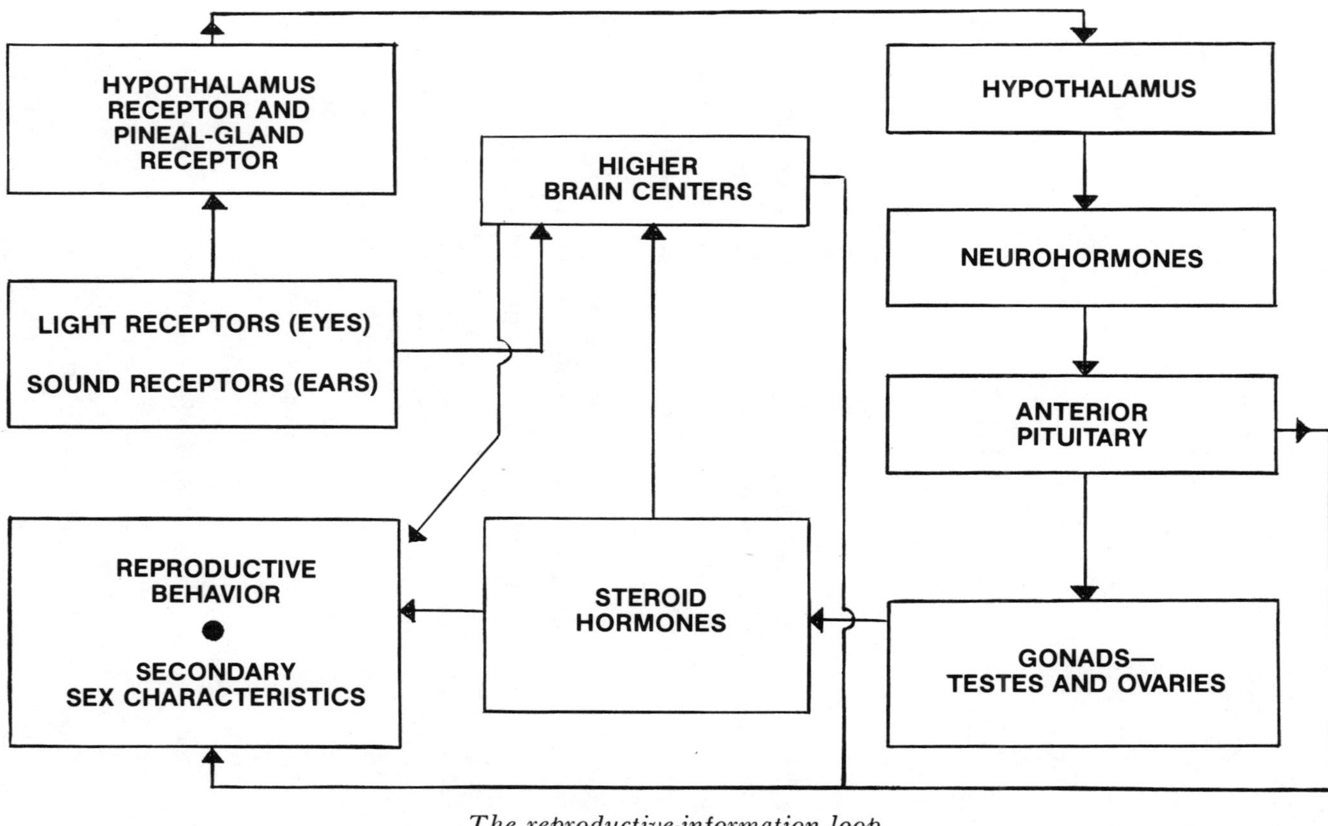

The reproductive-information loop

pituitary, and hypothalamus are involved in what we will call the reproductive-information loop.

A simplified information loop, modified from the work of D. S. Farner and R. A. Lewis in 1971, begins with light being controlled by the hypothalamus and the external sensory receptors (exteroceptors)—the eye, for instance—as shown in the diagram.

We see from this diagram why we call it a reproductive-information loop. Generally speaking, and keeping in mind the numerous experimental difficulties of decoding the exact system, the environmental-light information feeds to the higher brain centers from the eyes and also directly to the hypothalamus through the hypothalamus receptors, the nature of which is unknown. The hypothalamus in turn secretes the neurohormones that stimulate the anterior pituitary to activate the gonads (testes and ovaries). Once these organs are activated, they secrete the steroid hormones that enhance the secondary sex characteristics (and thus the maleness or femaleness) of the individual. Steroids also trigger the reproductive behavior of the bird. At the same time, feedback signals from the higher brain centers by way of the exteroceptors (eye and ear) initiate the sequence of events that lead to the display patterns necessary to successful copulation between the sexes. We might liken these display signals to a set of final stimulatory messages that push the bird across an inhibiting sexual threshold and into the copulatory and nesting patterns of the species.

Bird Communication

During early springtime most of the songs we hear originate with the male bird. What we hear is an intricate mechanism of survival that has evolved in the behavioral life of the species.

There is considerable evidence that male birds learn their complex song signals by imitating and learning the songs of adult birds. They also have the ability to improvise songs. The Oregon junco was found to have a wild type of song even though it was raised in a complete isolation from adult birds.

Songs coordinate most of the reproductive behavior of birds, so ornithologists classify them according to certain functions.

1. *Territorial songs* repel other birds of the same sex while at the same time they attract a mate. An example is the first calls of the meadowlark that one hears in country fields early in the spring. The male is warning away other male meadowlarks and "advertising" for a female partner.

2. *Signal songs* coordinate the activities of mated or courting and nesting birds. Signal songs are usually heard during the nesting season. An example is the "nest-changing" song of the male nightjar. The male and female nightjar must change on the nest with as little clumsiness as possible to avoid damaging the eggs. A male nightjar approaching the female on the nest utters a curious *quaw-ee, quaw-ee*. The female responds with soft *churrs*. They snuggle up against each other, churring and swaying their bodies, until suddenly the female flies off, leaving her mate to incubate the eggs. The *quaw-ees* and simple churring control movements in the interest of safety—the same function as traffic lights have.

3. *Female songs* are heard from some species of birds but are not as common as the territorial songs of the male. Both the male and female mockingbirds sing. The female answers the male territorial-advertising song.

The word "song" as used by bird behaviorists is a general term and there is much controversy as to the exact meaning. Many behaviorists believe the term should be restricted to the male territorial song alone. Niko Tinbergen, who received the Nobel Prize for his investigations into animal behavior, coined the term "advertising song." He believes that the male is advertising for a mate while at the same time threatening to repel any male rival.

Dr. M. Moynihan of the Smithsonian Institution has analyzed tropical American songbirds and points out that such a song definition may be inadequate. He defines a song as "any vocal pattern which, when uttered by one bird, usually repels other birds of the same sex and of the same species and attracts other birds, of the opposite sex." By this definition a song is any group of notes or sounds, from the complicated production of a mockingbird to the quite simple "loose quavering trill" of the Oregon junco, which keeps it all on one note.

It does not make any difference whether or not the male claims a territory. This latter definition seems to cover many of the songs that we hear in the spring. Most researchers believe that the main function of these song signals is to "stake out" a territory of sufficient size to ensure a good food supply for the nestlings, and also to locate a mate so as to ensure the propagation of the species. These complex song signals are received by the ear "detectors" and fed (as shown in the information loop) to the higher brain centers, and then to the hypothalamus, where the neurohormones trigger the appropriate behavior—and "the voice of the turtle is heard in the land."

C. B. Moffat and Territorialism

Song is of course taken in by the ear but there is another type of behavior that depends on the eye for its signal-reception. It is termed display behavior. In order to make any kind of sense out of display, a researcher must devote weeks or months to observing and recording the overall activity of birds in the field. This is one of the most enjoyable types of study and requires only a lot of patience, a pencil, notebook, and sometimes binoculars.

One of the first persons to notice that birds are territorial and that their message is conveyed by various forms of visual display was a newspaper man by the name of Charles B. Moffat.

He was not a professional zoologist but trained as a lawyer. Law did not suit his temperament so he left the bar and joined the staff of the Dublin *Daily Express*. He wrote lucid articles on natural history and became known as the "Bugman" to his colleagues, despite the fact that his main interest was birds and mammals. Moffat had an encyclopedic knowledge but was shy and modest. A colleague on the paper said of him, "To some of us he seemed to possess birdlike characteristics. Even in the way he entered the sub's room every evening there was something birdlike—at first poking his head round the door as if to explore that the coast was clear and then a dash forward and a pleasant 'good evening' to all." One of the staff once remarked that he would not be surprised if Moffat "flew in and lighted on the gas bracket."

Moffat was a member of the Dublin Naturalist Field Club and at one time or another was president, secretary, or treasurer of the Royal Irish Academy and the Zoological Society of Ireland. Professor J. S. Farley in *An Irish Beast Book* says, "Moffat was probably the finest field naturalist that Ireland ever produced and a perfect example of the theory that naturalists are born, not made. Not only had he acute powers of observation but the patience to watch and wait for events to take place." He was closely associated with the Irish Society for the Protection of Birds and was responsible for a bill passed in the Irish Senate in 1930 designed to protect them.

I am telling about Moffat at some length because I have often heard young people say that there are so many "protection laws" these days that it is hard for a young, inexperienced person to conduct any type of research on birds. That is decidedly not true, for any person might contribute a great deal to our understanding of birds by sitting under a tree, like the Indians I mentioned in the first chapter, and recording in his notebook exactly what he sees. This is precisely what all of the great naturalists did and it does not cost a penny, except of course for the pencil and notebook. This is how Moffat did things and is partly why he is so honored among ornithologists. He published his observations in the journal *Irish Naturalist*. He had noticed that certain species of birds "display" their colored breasts or other parts of their anatomy whenever a strange male wandered into their vicinity. He did not confine his descriptions of display behavior to birds alone, however, for in his paper on the Irish stoats (certain kinds of weasels) he wrote about how two stoats play: "One part of the play is particularly characteristic. Two stoats sit facing each other in kangaroo-like attitudes. Suddenly with a signal cry, both rush forward, spring into the air and there cannon against each other, then, falling to the ground, rush on and occupy each other's original places. . . ." This description sounds as if it might be a form of territorial display in mammals and is precisely what Moffat was the first, probably, to discover in birds.

Display Behavior

For one male bird to "understand" that another has staked out a territory for a nesting and feeding range, there must be some form of communication between the two individuals. The same holds for mating behavior. Visual display, like sound, is a form of such communication.

By the word "display" behaviorists generally mean "movements, postures, or sounds, of a specialized type, which have the capacity to initiate specific responses in another individual, usually of the same species." That is the definition of the great animal behaviorist and Nobel laureate, Dr. Konrad Lorenz. He calls such display signals "the language of animals."

Dr. Lorenz does not mean, of course, that birds and mammals actually speak and reason with each other as do humans, but rather that the display signals are inborn. The elaborate signals are understood between the communicating individuals without any learning process beforehand; they are automatic and nearly mechanical in nature. Professor Edward Wilson coined the word "sociobiology" to refer to behavior that is inherited. Although inherited social behavior among lower creatures such as birds is well accepted in scientific circles, inherited social behavior in human beings is not widely accepted at all. Wilson believes it occurs in humans as well as all other living creatures and his ideas have become very controversial among modern biologists.

Bird display signals are quite involved and difficult to interpret. We know very little about the visual signals that lead to the pair bond in most species. Ornithologists classify the display signals according to whether they involve threat or appeasement, which is the opposite of threat. Threat display usually takes place between males and involves the defense of the home territory. Appeasement is the type that cancels out the threat and brings the male and female bird together. If it leads to copulation it is called epigamic display. It occurs only during courtship.

Any display that takes place between nesting pairs is called postnuptial display. It is believed that this type of display keeps the pair together during the nesting season.

The Ceremonial Gannet

Stylized postures and body movement, including certain color patterns, are a visual language and carry clear information between the displaying bird and observing bird. In many cases the movements are so stylized and complex that behaviorists call them dances. Whenever colored parts are used in such dances they are positioned so that they are clearly visible to the observing bird. Such displays usually involve the fluffing out of colored feathers or the lowering or raising of a crest. The beau-

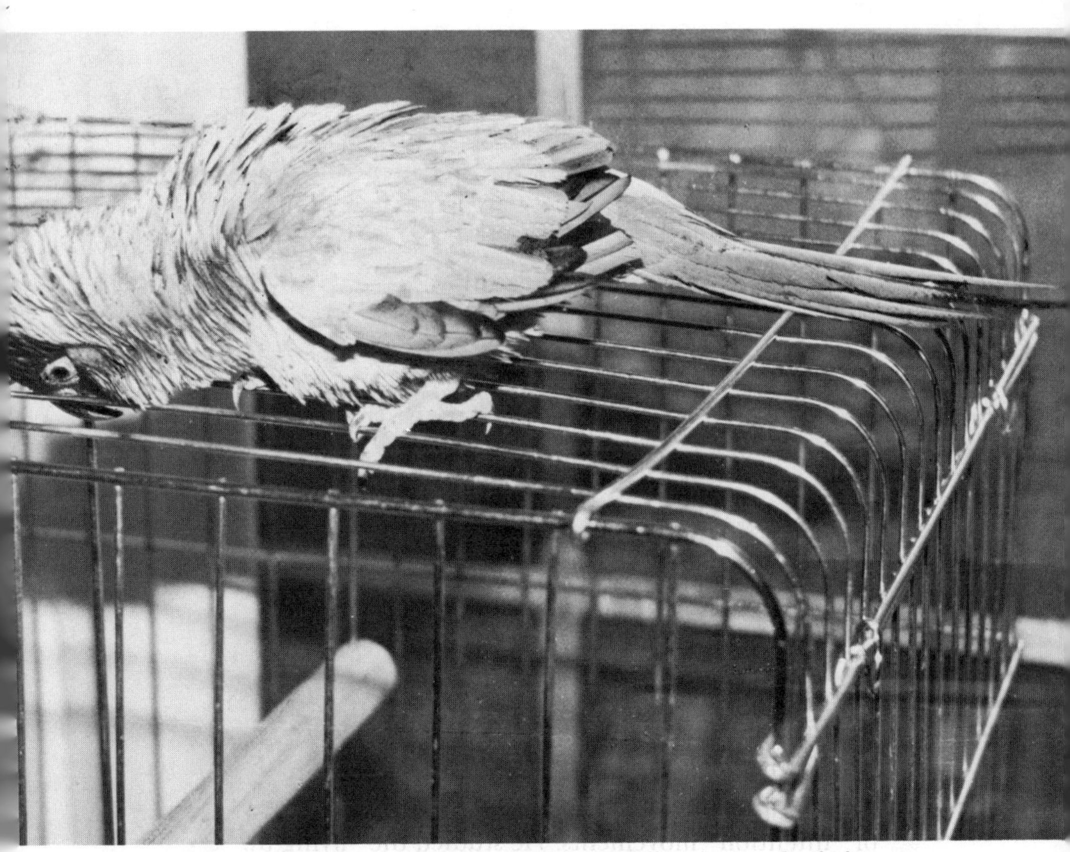

This Nanday conure, a type of parrot, is standing in a submissive display posture on the top of its cage. Birds often assume this position before a bird that is dominant. This conure has been imprinted by the author, who in effect takes the place of a companion conure.

tiful, brightly colored plumage of the male bird of paradise is an excellent example of display plumage.

Since display behavior synchronizes the activities of the sexes through the sense organs and brain centers it must be understood as a series of actions and following reactions on the part of two actors, male and female. Behaviorists call the bird that initiates the display (usually the male) the actor, and the bird that responds (usually the female) the reactor. As mentioned before, behavior that suppresses aggressiveness on the part of the actor, and escape on the part of the reacting bird, is called appeasement display. After appeasement and copulation the birds must remain together throughout the nesting period. Postnuptial displays, which keep the pair bond strong, are sometimes referred to as ceremonial displays.

The gannet is a large sea bird that nests in colonies on coastal cliffs. It is common in the northern American waters and also in Europe. In fact, the North Sea was known by early sailors as "the Gannets' Bath."

Edward Armstrong in his classic book *Bird Display and Behavior* introduced his subject thus: "I will ask the reader to imagine himself transported on the swift wings of fancy to the midst of a gannetry. There let us notice some of the birds' curious customs and mannerisms. Later we shall reflect on our observations at leisure, making them the starting point for a discussion of various problems of bird life and the elucidation of their significance."

In the book the author describes the "flying-up" ceremonies of that species. He calls these postnuptial activities initiating or "intention" movements. He studied the "flying-up" ceremony in relation to the birds' structure, their social behavior, and their method of incubation. Dr. Armstrong arrived at a gannet colony during the breeding season, and so did not believe the posturing of the nesting birds to be epigamic display; but he cautions that it is difficult at such periods to separate the two types, despite such ornithologist-made terminology as "epigamic" and "postnuptial." As he points out, the one obviously overlaps the other.

The "flying-up" ceremony begins with "billing" behavior. As a gannet returns to its mate they "face each other, stretch up their heads with beaks pointing skywards, partially open their wings, and waggle their heads energetically so that the beaks clatter as in a kind of fencing or scrape together as if they were being whetted one upon the other." This entire ceremonial greeting is accompanied by a hoarse throaty shout sounding somewhat like *rrah, rrah, rrrraaaah.* The cry is often transmitted from one group of birds to another until the entire cliffside is ringing with the strident chorus.

In May 1978 I witnessed this same scene from a fishing boat off Skellig Island, Ireland. It brought to mind the ancient song from the *Perils of the Seafarer* and quoted in Armstrong's book:

> *At times swan's song*
> *I made to me for pastime,*
> *the gannets cry,*
> *and the hu-'ilpse's cry*
>
> CODEX EXONIENSIS

Across those waves came the cries, songs, and visions of millions of years of evolution. I was witness to the ultimate consequence of what this entire book is about—a series of organic events, involving sound and vision, and leading to the survival of species through the mysterious function that we call reproduction.

12/ MIGRATING AND SURVIVING

In 1963 a new species of old and extinct bird appeared in the taxonomic, or classification, literature of ornithology. The genus was also new. By any standards the description of the bird does not make for exciting reading. It was written by Dr. Pierce Brodkorb of the University of Florida. The description reads like this:

> *Distal portion of the right tarsometatarsus, University of Florida no. 4108, with referred right pedal phalanx 1, digit III, UF 4109. From late Pleistocene deposits of Santa Fe River . . .*

The rest is detailed description of the leg bones of a giant flightless bird found in the sandy bed of the Santa Fe River in Florida, a few miles north of my house in 1961. Although the description is dull, what it conjures up has all the drama of a well written mystery story.

Titanis walleri apparently roamed over the flatlands of Florida during the early Pleistocene, a mere two million years ago. That was during the evolution of the human species and by the geologic time calendar not so long ago.

According to Dr. Brodkorb's own words, it was a "bird of tremendous size, larger than the African ostrich and more than twice the size of the South American rhea." What happened to this animal? Was it wiped out by a strange disease? Did the ecology of Florida change so that it no longer "fitted" its environment? Were the nesting habitats flooded? Did some ancient predator like the saber-toothed tiger kill off the entire population? Why are there no more *Titanis walleri* racing across the grassy lands of Kissimmee prairie in central Florida? We will probably never know but we may well suspect that the true reason was the evolutionary rise, at the same period, of the most heartless and efficient predator of all ages—the human species. And therein lies the tale of disaster for many living creatures at an ever-increasing rate.

Moas and Bogs

The giant moa of New Zealand is a case in point. Actually, there were seven genera and about 25 species of moa; and only a few of them were real giants. They ranged in size from the 12-foot-tall giant moa to pigmy moas the size of a turkey. The bogs and swamps of New Zealand abound with the bones of various species of moas.

New Zealand scientists believe that there is little doubt that all of the moa genera and even most of the species survived into the human period. On South Island there are a considerable number of human culture sites where moa bones have been discovered. Human beings hunted the moa and its eggs for food, and the fact that the giant moa was known even to us moderns indicates that they probably hunted it into extinction. Humans, however, were probably the last causal factor in the series of ecological changes that led to the extinction.

What is known in New Zealand as the postglacial climatic optimum reached its peak at about 5000 to 4000 B.C. As the glaciers retreated, the climatic optimum was marked by long wet periods accompanied by a huge extension of forests.

As the forest spread, the grassland areas became more and more restricted; swamps and bogs, which are natural traps for the long-legged, heavy moas, became abundant. According to C. J. R. Robertson, 40 or 50 centuries of such miring would account for the vast accumulation of moa bones found in such swamps.

Without a doubt the same fate overtook the giant flightless *Titanis walleri* of Florida. The climate changed and forest and swampland extended over Florida, leaving only the central grassy area of the Kissimmee prairie. Large flightless birds driven into such a restricted "ecological corner" could have been easily hunted to extinction by early North American humans. Might these marvelous birds have survived had they been winged and able to migrate to other areas?

Migration

From earliest times, human beings have known that during the cold winter periods, many birds leave for warmer climates. Frederick II, Emperor of the Holy Roman Empire during the thirteenth century, studied bird migration by having his slaves capture and tie colored bands to migrating birds. His soldiers stationed in North Africa would recapture the birds and send the information on species and times back to him in Italy. And from that time to this, more and more information about migration has been amassed. Researchers are working on the problems daily, studying varying aspects—not only migration areas and route-finding by birds, but the physiology of ear, eye, and brain. Display signals act mainly through the eye and ear sensory mechanisms in the reproductive-information loop shown in the preceding chapter, but this does not seem to hold for preflight restlessness (scientifically called zugunruhe).

Modern researchers (using cruel methods that I neither approve of nor would prescribe) have removed the eyes of birds

Canada geese on Payne's Prairie in Florida. Most such geese spend the winter in the South but migrate again to northern Canada for breeding.

and replaced them with quartz tubes that directed the light to the deep regions of the brain by bypassing the pituitary gland. They found that this direct pathway of stimulation to the hypothalamus, through skull and brain tissue, may in the wild be even more important as a stimulus, in the case of such phenomena as the preflight restlessness, than the light pathway through the eye.

This experiment shows clearly that body tissue—including the thin bone of which the skull is made—is translucent and passes the light radiation directly to the hypothalamus. A point that needs to be made about the methods used in this experiment, however, is that these researchers could have conducted valid experiments by hooding the birds as falconers do, and leaving an open area in the hood at the top, above the hypothalamus, for the light to penetrate the brain tissue. When the experiment was over the birds could have been released—something one cannot do with blinded birds, of course.

Distinguishing Dispersal from Migration

Ornithologists define migration as periodic movement of birds from one region to another in specific directions. There are many excellent books on the subject. Bird-banding, as begun by Frederick II, has reached such a state of the art that most of the flyways of North American and European bird species are well known. There are, however, large gaps in our knowledge of what ornithologists call dispersal.

Dispersal is a phenomenon of bird movement that is often confused with migration. Again, it was Frederick II who first distinguished between the two. He pointed out in his book on falconry that falcons, and the herons he hunted with their aid, dispersed from their nesting places and spread out fanwise over large areas of the countryside.

One of my favorite songbirds is the tufted titmouse. I think

I love the twinkle-eyed little bird because, no matter what the time of year, even during the times of deepest snow, its cheery call always fills the still woodland air. Birds such as this one are considered nonmigratory, because they remain at the same latitudes that are their normal nesting areas. Such birds, although they have evolved to survive the rigors of winter, must still disperse and fan out from the nesting area in search of food. In winter it is always the same birds we see around the feeder, birds such as these titmice, woodpeckers, chickadees, and cardinals. All of these species are insect-feeders during the summer, but do quite well on seed when forced to switch their diet. They are able to fit into summer and winter environments about equally well.

Dr. Joseph Hickey divides dispersal movements into three different kinds: periodic, annual, and irregular, or accidental. The snowy owl is an example of periodic dispersal. When its food supply, small rodents called lemmings, becomes scarce in the north, as happens periodically, the owl disperses far south of its normal range in search of food. Annual dispersal occurs when birds wander away in search of nourishment after the young are fledged. Tufted titmice, falcons, herons, and cardinals are examples of species that disperse annually. In actual fact, almost all birds fall into this rather broad class. Irregular dispersal occurs when birds are accidentally driven by high winds or gales on long, sustained flights to new regions. The African cattle egret, now very common in the South, is believed to have reached Florida in this way.

Most dispersal movements are temporary, because the birds move through a new region, stopping to feed now and then but not adopting it as a new residence. The cattle egret, although it arrived accidentally in the United States, found an empty ecological food niche to occupy. It feeds on insects stirred up by grazing livestock. No native heron has such a feeding habit. Such an accidentally induced but permanent dispersal, in which a

species expands its home range, is called "spread" by some migration researchers but not all.

Usually accidental dispersal does not favor the survival of the individual bird, as is demonstrated by what ornithologists call "rare" sightings. These are birds far from their home ranges. Many such sightings occur at lighthouses—which I consider one of the most fascinating of migration phenomena. Researchers appear to have no valid explanation for this, but I offer one that seems to me very convincing.

TV Towers and Lighthouses

Scattered around England, northern and southern Ireland, Wales, and Scotland are several lookout posts described in a book titled *Bird Observatories in Britain and Ireland*. Several of these are located on coastal islands with lighthouses on them. One called Bardsey is off the coast of northern Wales. Over the years Bardsey has had a considerable number of rare sightings involving North American birds. Species over the past 20 years have included a summer tanager, yellow warbler, blackpoll warbler, white-throated sparrow, and several gray-cheeked thrushes. Note that most of these rare accidental migrants are insectivorous songbirds.

The fact that insects of all types fly to lights is well documented and easily observed at one's own front-porch light. That "fact," however, is not exactly a truth; if one takes the time to count the insects at a street light or other outdoor bulb, it will be seen that most of the insects are not at or on the bulb iself but in the airspace around it. One might want to call that statement a splitting of hairs, but from a scientific viewpoint it is very important.

Cattle egrets, originally from Africa, probably arrived in the United States as a result of being blown here by gales. Finding that they could fit in successfully, they have remained.

I have been able to show in a series of papers reporting various experiments that molecules of various scents (sometimes from nearby trees) plus molecules of the air itself—mainly of oxygen, hydrogen, carbon dioxide, and water—around the light bulb are the real attractive force. The insect antenna really is an *antenna*; it resonates to short coherent (radiolike) waves from the little molecular oscillators surrounding the light bulb. It is a very complex phenomenon, but in simplified terms, the energy from the visible radiation given off by the light is "trapped" and "scattered" by the air molecules, which then reunite the waves as weak infrared radiation. The insects are not attracted to the visible light at all but to the area of coherent scattered infrared waves surrounding the bulb; it is the insect's antennae and not its eyes that are the sensors for this attraction.

In 1968 I put insect traps that emitted ultraviolet radiation high up on the side of a 1200-foot television tower; these attractant blacklight radiators were in the bottom of tin cones and so could be seen only from above the tower. I trapped over 15,000 large specimens, which I counted, but could not count the millions upon millions of tiny insects that clogged my traps, especially during the fall migratory season.

Most birds, though not all, killed at lighthouses are insect-eaters, such as warblers. Migratory birds use a lot of energy during long, sustained flights and need food badly. They are attracted to the insects, which are themselves attracted to the air-space around the light.

One might ask: Why, then, do we not see the insects at the lighthouse? The answer is that during peak insect migration nights in the fall one does see the very large insects. Dr. C. B. Williams, in his book *Insect Migration,* lists over 120 species of lepidoptera alone captured at lighthouses, in the form of light-ships, far out at sea. Unfortunately, what are seen and captured by entomologists are only the large, obvious species. No one bothers about the millions upon millions of minute insects, like those trapped in the cover of a porch light, that fill the airspace around the lights. These tiny creatures cannot be seen at all in the powerful glare of a lighthouse bulb and are quite far out in the airspace around it.

Most birds feeding on insects around lights are not killed, but a few follow in too close to the powerful glare, become disoriented, and crash against the protective cover of the light or against the building itself.

Wavelengths and Power

The next question that is likely to arise in one's mind is: What about the small red warning lights on television towers—surely they are not strong enough to attract such masses of insects? True, they could not equal a 500,000-watt lighthouse lamp, but in terms of power the TV antenna may often radiate 500,000 watts of radio-wave energy into the airspace—enough to cook a steak! The simple fact is, as any laser scientist will attest, TV radio waves will also stimulate little molecules to give off coherent wavelengths in the same manner as a bulb's light waves or an ultraviolet source. The radio- and light-wavelength energy going into the airspace is called "pumping waves." The actual wavelengths of such a "pumping system" are not critical as long as they are *powerful* enough. The coherent waves that are produced from air molecules under stimulation by radio or visible or ultraviolet light energy are in the far or intermediate infrared portion of the spectrum—usually in the far infrared, toward the microwave end of the spectrum. This is where the insect antenna responds effectively to infrared wavelengths.

During my 1200-foot-tower experiment I caught 15 ovenbirds (a species of insect-eating warbler) in my blacklight traps. Since the traps could be seen only from above and did not radiate downward, it was obvious that the little ovenbirds were above the traps and followed the insects down the cone and into the trap jar, where they died. The radiation from these traps was of invisible near-ultraviolet, which birds cannot see but which is visible to insects. Birds would not be attracted to radiation they cannot even sense. Ultraviolet is a powerful pumping radiation for airspace molecules, so insects are attracted to such traps. Since there was no visible light to blind the ovenbirds, it was a deliberate flight after insects down into the traps.

On certain nights when I went up the tower in darkness I could see only a few large moths at my traps. Smaller species of insects were visible with a flashlight and only close by. At the 1200-foot level, which I could not approach because of the dangerous TV radiation, I saw a few large moths. During daylight, wasps were attracted to the dipole radiator. Most of the insects attracted to the TV tower were at the very top levels around the dipole and in the airspace above my traps at their own lower level.*

In addition to crashes against lighthouses due to disorientation by the light, birds die in TV-tower crashes. Research conducted by the Tall Timbers Research Station at Tallahassee, Florida, on a nearby TV tower demonstrated that most of the birds were collected on the ground around the tower below guy wires that help support it. A thousand-foot fall by a bird hitting such an invisible wire in the dark is certain death.

In this chapter we can see the truth of the statement by the great naturalist John Muir, "Everything is connected to everything else." The light waves or TV radio waves are connected to the airspace molecules, which in turn emit infrared waves, that in turn attract millions of insects to the airspace around the radiators, which in turn attract the hungry birds, which in turn leads to death for some.

* After writing this chapter, I built a system that emitted three-centimeter microwaves and tried it out on several different kinds of organic molecules, which I blew through the waves. In every case the organic molecules were stimulated by the microwaves to emit coherent infrared radiation, in various parts of the infrared spectrum.

Thus does modern human technology affect the very survival of birds. I have all my life been charmed and preoccupied by the mysteries of nature, and especially by the effect of technical advancements on the living things of our earth. We will see presently how, under certain unusual conditions, migration may place birds at an even greater disadvantage than do a relatively few TV towers and lighthouses.

13/ SAVING OUR BIRDS

Whooping cranes are large white birds with a black and red skullcap. They are named for their powerful whooping cry, which if often given in duet as they perform their remarkable courtship dance.

I first saw whooping cranes in 1942 during World War II when I was living in San Antonio, Texas. I was 18 years old, traveling with a famous wildlife artist. We had driven to the town of Aransas Pass on the Gulf Coast of Texas. Six years before, in 1936, a small band of overwintering whoopers had been discovered on the large grassy and marshy prairie that made up the Blackjack Peninsula.

Mrs. Hager, the famous "bird lady" of Aransas Pass, awoke us at our motor court at 5 A.M. and by 5:30 we were out on the prairie watching the stately creatures perform their early-spring courtship dance.

It was a day I shall never forget as long as I live. And when Mrs. Hager told us that there were only 15 whoopers left in the entire world, I was saddened that this beautiful creature might soon join the passenger pigeon and the Carolina conure on the list of extinct species.

The literature of today gives the impression that the U. S.

Fish and Wildlife Service rushed right out and bought up the Aransas Pass area to save these last 15 cranes. Of course it didn't, because that's not the way life really works; but eventually they did manage to set up a 47,000-acre wildlife refuge in the area. By 1978 the Aransas population had grown to 71 individuals, despite the fact that at that time it was safe only on its far northern breeding grounds. It was known to migrate across the entire United States and into northern Canada but nobody knew where the species nested. The long migration flight takes place twice each year to and from the nesting grounds. In 1945 the whoopers' nesting area was discovered in Wood Buffalo National Park, which lies in Alberta and the Northwest Territories of Canada; thus their migration line lies over 1500 miles of western United States and Canada. The birds cross Texas, Oklahoma, Nebraska, South and North Dakota, a corner of Montana, Saskatchewan, and Alberta to reach their nesting marsh in Wood Buffalo Park. This is obviously a journey full of danger for these large conspicuous birds.

A Whooper Test

Several years ago the Fish and Wildlife Service funded an interesting survival project. It was believed that, while the nesting population in Canada was in a fairly safe area, the Aransas refuge did not seem ideal. Not only might a seasonal hurrican wipe out the small population, but even more to the point, an oil spill or toxic chemical leak from the huge traffic of barges that pass through the coastal waterway is a constant threat.

Biologists under the leadership of Dr. Ray C. Erickson decided to start a flock from hand-reared cranes at Grey Lake National Wildlife Refuge in southeastern Idaho. There are all kinds of breeding problems, and even if those problems are solved and a good breeding stock established, the question still remains: will hand-reared birds migrate south to the Bosque del Apache National Wildlife Refuge in Arizona, where they would be well protected?

The reason it is believed that they will migrate between those two refuges is that those areas are the northern nesting and southern wintering area of the sandhill crane, a relative of the whooper. The whooper eggs are being foster-parented by sandhill cranes by researchers' placing the eggs in viable sandhill crane nests. If the breeding experiment succeeds, it is hoped that the foster-reared whoopers will follow their sandhill crane cousins back and forth and finally develop a colony of their own. If the whoopers take to the sandhill foster flock flyway, from Idaho to Arizona, it will have cut the migration distance of the species to one-third its former distance. I consider this one of the best possibilities to save the species, because long experience with large conspicuous birds of prey leads me to believe that gunfire is the greatest hazard that such species face. The shorter the distance that large migratory birds have to fly, the greater the chance of surviving the great American sky-directed artillery barrage.

Firearms, Cars, and Insecticides

Eagles and cranes are highly conspicuous but I do not mean to imply that gunfire alone brought these birds to near-extinction. One of the greatest factors is of course habitat destruction, especially in the case of cranes whose nesting marshes in Canada and the northern United States were drained and turned into agricultural land. However, it is ridiculous to maintain that gunfire alone cannot bring a species to extinction. It did a very good job on the buffalo long before the prairie vanished into wheat fields.

Despite the fact that birds fly and are capable of quick escape, the invention of firearms nullified that advantage. When

The American bald eagle is close to extinction, largely due to destruction of habitats and to hunting. This adult was shot in the wing and rehabilitated by the author, to be released later back into the wild.

I was a boy growing up, bird-lovers' societies were forever blaming farmers for shooting hawks—"chicken hawks," as all birds of prey were called in those days. The fact was that farmers simply didn't have time to run about their fields banging away at soaring birds of prey. I am an agricultural scientist and have been around farmers for 30 or more years. Farmers don't like guns on their property. The fact is that most indiscriminate and illegal shooting, *not hunting*, is done by urban dwellers who buy a gun, get in their car, drive along the highway or into the field, and shoot at anything that moves. There are 20 times as many guns around today as during the Great Depression, when I grew up, and they are mostly in the hands of thoroughly irresponsible people.

There are probably only 100 or fewer bald-eagle nesting sites in Florida, yet in 1978 the Florida Audubon Society reported nine bald eagles picked up with gunshot wounds. I helped rehabilitate two of them. Since most eagle nests usually contain one fledgling per nest, this is approximately 10 per cent of the yearly production of bald eagles (our national emblem) in Florida. In all probability, nine or ten more are killed outright and not picked up wounded, so 20 per cent could be closer to the truth.

An automobile is a deadly weapon and so is licensed to insure proper use. Today, in order to own a hawk or hunting falcon, the falconer must pass a rigid test and be licensed; yet all the falconers that I ever knew are practicing conservationists. Why on earth ownership of firearms does not require equal demonstration of responsibility seems beyond all comprehension.

Many scientists believe that overkill hunting, for the restaurant trade, was the main cause of extinction of the passenger pigeon. There were obviously other contributing causes, some no doubt ecological, that we will never discover. Once a species becomes extinct, there is no method by which one may further study the behavior in order to account for the causes. Extinction is an accomplished fact!

The Needless Deaths of Peregrines

In almost every case in which a species has been placed on the endangered species list, researchers invariably have found that a number of factors were interrelated. The eastern subspecies of the peregrine falcon (*Falco peregrinus anatum*) is extinct and most evidence points to a combination of human interference at nesting sites, destruction of eggs due to the effect of DDT on the reproductive metabolism of the bird (eggshell thinning), and shooting.

It is now a fact well documented from field and laboratory studies that excessive use of DDT and other insecticides causes the eggshells of certain species of birds to be thinned and thus easily broken in the nest. Although the metabolic pathway of this effect is still not known, the amount of DDT and other chlorinated hydrocarbons put into the environment was enough to account for the phenomenon.

Certainly two other factors were involved in the extinction of the eastern peregrine. From 1940 onward I was well acquainted with eight or so nesting eyries of peregrine falcons in New York State and Vermont. By the early 1960s most of these had disappeared. During this period, when DDT was building up to intolerable levels in the soil of the eastern United States, so also was human interference at falcon eyries—and I do not mean from falconers, as some bird-lovers maintain.

I am a falconer myself and during that period I took only one bird from those nests. Most American falconers are quite satisfied to own and train a single bird, and, unless they have bad luck, they keep it for a considerable number of years. Of six nesting sites that I had known about in the 1940s, before and after World War II, and located in Vermont, all had been taken over by the mid-1960s and the woods nearby cleared for motor campers. In three out of the six sites, the camping area was directly below or above the nesting cliff. At one site (Whiteface Mountain) a trail was cut down a steep decline within 50 feet of the nesting ledge. At another nesting site, that has been known to ornithologists for over a century on the face of a cliff in the Helderberg Mountains in New York, the falcons disappeared in the early 1940s. At that time cars became more numerous on the overlook that had been cleared above the cliffs. Beer bottles, paper, and cans were thrown over the cliff face.

After World War II, Robert Bauer, who was with the New York State Game Department, and I banded the birds on the Whiteface Mountain eyrie. Before the fall was over we had returns on two of the four nestlings. One had been shot near Jamestown, New York, and the other near the town of Falcon (!), New York. DDT may well have hastened the extinction of the eastern peregrine but in my opinion two other factors, human interference and shooting, were equally involved in the sudden decline and extinction of *Falco peregrinus anatum*.

If our rare and endangered birds are to survive, attention must be paid to not only ecological and habitat problems but also to some sort of workable control of firearms so that responsible hunting is encouraged and indiscriminate and irresponsible shooting stopped. There are more gun-owners than ever, and the educational campaigns of the conservation societies have done little to discourage the irrational killing of our feathered creatures. It must be stopped.

AN AFTERWORD: BIRDS AS PETS

The human population is slowly closing in on the wild creatures of the world in much the way the advancing settlers almost wiped out the buffalos of our own great plains. There are believed to be about 11 species of parrots either extinct or on the very edge of extinction, thanks to human activities within the past quarter-century. Most of these species are threatened by the extreme habitat destruction taking place in the forests of the world. We are fast destroying and polluting our environment. Charles Walters, Jr., in *Acres USA,* points out that in Marco Polo's day fully 85 per cent of China was forested. In 1948, when I was hiking around China, I saw no forests but only bare, eroded hills. A tree was a rarity except in cities.

Humanity was born of nature and seems to have an inherent desire to be close to wild creatures just as it has an inborn drive to hunt. These are probably evolved traits of the 4,000,000-year-old creature we label *Homo sapiens,* yet the civilizing of our species demands that we use our intellect to control these "needs," since they no longer fit our way of life.

I am not suggesting that we attempt to eliminate hunting or pet-keeping, but rather that we assure the continuance of wild populations by fitting our evolutionary traits to our modern technological way of life. For a good review of our greedy and inhuman (perhaps I should say "human") treatment of creatures trapped for the pet trade, read the article Jean-Yves Domalain's "Confessions of an Animal Trafficker" (Suggested Reading). Edward R. Ricciuti in "The Bird Lovers," also in Suggested Reading, points out that the rarer and more threatened the species, the higher the profit to the animal trader. That rare golden-shouldered parrot smuggled out of Australia may bring $10,000 on the open pet market. A peregrine falcon from Iran sold for $25,000 in Saudi Arabia. Ten thousand birds of prey are imported by Hong Kong for sale in mainland China to be ground up for use as folk medicine to insure virility.

Conservationists who believe that we can outlaw pet-keeping and hunting not only do not understand the evolution of the human species but live in a dream world. They believe that they can take out of us what evolution spent 4,000,000 years putting into our makeup. What protectionists can do is to require that dealers in firearms and pets help educate the coming generations in the proper use of the goods that they market. Shops, in the case of pets, should sell only species that are common and not endangered. They should be species that fit well into human urban environments, due to their behavioral characteristics and size. Wildcats and golden eagles do not fit; starlings, cockatiels, Nanday conures, and dogs do fit. If you want to own a parrot, don't buy a cockatoo or macaw, all species of which are threatened. Buy instead a cockatiel or a Nanday conure, which are easily trained, common, and need not even be imported, for they are easily reared in captivity.

All native American birds are protected except English sparrows and starlings, and therein lies a paradox. A starling makes a far better cage bird than does a hill myna—for which the native trapper in Asia might receive from fifty cents to five dollars but which is sold in pet stores here for $300 or $400.

The starling is a beautiful bird when viewed up close, and can whistle just about anything. If taken young from the nest,

Human beings have evolved for some two to four million years (depending partly on one's definition) but the golden eagle has had 50 million years of evolution. Thanks to us, eagles are in danger of ceasing to exist. Will we wipe them out?

it is easy to imprint and train, and also to feed.

Mr. Ricciuti believes that about 6,000,000 songbirds are hunted and killed illegally in southern Europe. The Royal Society for Bird Protection estimates that 5,500,000 birds, mostly rare or endangered, enter the pet trade every year. Before you, as a bird lover, contribute to such activity, stop and think carefully about what bird you choose for a pet. When you buy a wild creature, remember that its life is totally in your hands.

GLOSSARY

auditory canal—The canal leading from the opening of the external ear to the eardrum.

auricular patch—A patch of feathers covering the opening of a bird's auditory canal.

barbicels—Small hairlike processes resembling fringes on barbules, that hold the barbules together.

barbs—Branches that extend from the main shaft of a feather.

barbules—Small filamentlike branches that extend to the sides from the barbs of a feather; they help to hold the barbs together.

biofeedback—A technique that enables a person to control automatic bodily processes.

blacklight—Invisible light; specifically, infrared and ultraviolet.

bronchus (pl., *bronchi*)—One of a pair of breathing tubes that enter the lungs.

carrion—The dead and rotting flesh of an animal.

cerebellum—A hind part of the brain that is especially concerned with muscle coordination and keeping balance.

cerebrum—The upper forebrain of a vertebrate; the seat of thinking processes.

cloaca (pl., *cloacae*)—The chamber in birds, reptiles, and amphibians into which the intestinal, urinary, and reproductive ducts empty; in the female it also acts to receive sperm from the male during copulation.

coherent radiation—Visible light or infrared that has a very narrow frequency range, in that particular resembling a radio station's wavelength.

cone cell—A retinal cell that responds to color and strong light.

copulation—The sexual joining of a male and female for the transfer of sperm.

coracoid bone—A bird's breast bone.

cornea—The outer, transparent part of the eyeball.

crop—An expanded part of the digestive tract in many birds, for food storage and preliminary food-grinding.

dimorphism—A difference (as of color, size, or form) between two individuals, or kinds of individuals, that one would expect to be identical or nearly so.

dispersal—Spreading out of birds in many directions over large areas.

display—Any sort of signal, such as a song, call, posture, or other movement of the body that initiates and synchronizes reproductive activities. Some display is used for "advertising" occupation of a territory, for defending territory, or for helping to maintain a pair bond.

down—Small, soft feathers lacking a vein.

ecological niche—All the factors (such as climate, sources of food and water, natural enemies, etc.) that control a species' way of living; also, a habitat that supplies the right factors for the needs of a given species.

endocrine—A gland that secretes a hormone and feeds it to the bloodstream; also (adj.), relating to such a gland.

enzyme—A body chemical that promotes chemical reactions in a physiological process; for example, pepsin is an enzyme that aids digestion of proteins.

esophagus—The throat; in birds it leads into the crop.

flyway—A geographic route used by birds to migrate between a wintering and a breeding area.

fovea—A part of the retina that has especially great resolving power.

gizzard—The second stomach of a bird; it serves mainly for grinding food and also as a storage sac.

hamuli—Hooked tips of barbicels that hold the barbicels of a feather together.

hormone—A gland-produced chemical that triggers or regulates a bodily process.

hypothalamus—A part of the brain within the limbic system.

infrared—Invisible radiation of longer wavelength than the red part of the electromagnetic spectrum; though all bodies radiate some heat at any temperature above absolute zero, those emitting infrared can warm an object to the greatest extent.

keratin—A protein that makes up the main substance of hair, nails, claws, and hoofs.

limbic system—An area of brain tissue concerned with emotions, drives (urges), and vital physical functions.

metabolism—The sum of chemical processes that build up protoplasm and release energy from food for bodily use.

migration—Movement of birds from one region to another, usually periodic.

neuron—A nerve cell.

niche—See *ecological niche.*

olfaction—The sense of smell.

pair bond—The urge of certain mated animals to stay together, which benefits offspring.

photoperiodism—The capacity of a bird or other organism to respond to light and dark periods of various lengths.

pituitary—A gland in the brain that has a major function of directing other glands and many physiological processes.

primaries—Outer wing feathers of a bird.

pygostyle—The last bone of a bird's backbone; tail feathers attach to it.

resolving power—The ability of eyes to see fine detail at a distance.

retina—The light-sensitive membrane at the back of an eye.

rod (rod cell)—A retinal cell that responds to faint light but not to color.

secondaries—The inside feathers of a bird's wing; they attach to the arm (shoulder to elbow) of the wing.

sensor—A sense organ, such as an eye, a heat-receptor (Pacinian corpuscle) in the skin, or a taste bud on the tongue.

spermatogenesis—The creation of sperm by the body of a male animal.

tarsus—The lower portion of a bird's leg.

territory—An area including the nesting range and hunting range, held and defended by a male animal.

testes—Male glands that produce sperm.

trachea—The main breathing tube.

tracts (feather tracts)—Restricted areas, usually lines, along which a bird's feathers grow.

ultraviolet—Invisible radiation of shorter wavelength than the violet part of the electromagnetic spectrum.

vane—The part of a feather on either side of the shaft that is formed into a web.

wing loading—The amount of lift per square foot of wing needed to keep a bird or other flying object in the air.

zugunruhe—The premigratory restlessness of birds.

SUGGESTED READING

Books

Ardrey, Robert, *The Territorial Imperative* (Atheneum, New York, 1966)

Armstrong, Edward A., *Bird Display and Behavior* (Oxford University Press, New York, 1947)

Baldwin, S. P., and S. C. Kendeigh, *Physiology of the Temperature of Birds* (Scientific Publication of the Cleveland Museum of Natural History; Cleveland, Ohio, 1932)

Bent, Arthur Cleveland, *Life Histories of North American Birds* (Dover, New York, 1919-1958). These are reprints of the original series of works issued over a period of years by the Smithsonian Institution, through the Government Printing Office, Washington, D.C.

Berger, Andrew J., *Bird Study* (John Wiley & Sons, New York, 1961)

Brown, Leslie, *Eagles of the World* (David & Charles, London, 1976)

Brown, Philip, *Birds in the Balance* (October House, New York, 1966)

Callahan, Philip S., *The Evolution of Insects* (Holiday House, New York, 1972)

———, *The Magnificent Birds of Prey* (Holiday House, New York, 1974)

———, *Tuning in to Nature*. (Devin-Adair, Old Greenwich, Conn., 1975)

Dorset, Jean, *The Life of Birds,* vols. 1 and 2 (Columbia University Press, New York, 1971)

Durman, Roger, *Bird Observatories in Britain and Ireland* (T. and A. D. Poyser, Berkhamsted, England, 1976)

Farmer, Donald S., ed., *The Breeding Biology of Birds* (National Academy of Science, Washington, D.C., 1973)

Fisher, James, and Roger Tory Peterson, *The World of Birds* (Crescent Books, Crown Publ., New York, 1977)

Freedman, Russell, and James E. Morriss, *How Animals Learn* (Holiday House, New York, 1969)

———, *The Brains of Animals and Man* (Holiday House, New York, 1972)

Hickey, Joseph J., *A Guide to Bird Watching* (Oxford University Press, New York, 1943)

May, John Bichard, *Natural History of the Birds of Eastern and Central North America* (Houghton Mifflin, Boston, 1939)

McLuhan, T. C., compiler, *Touch the Earth: A Self-Portrait of Indian Existence* (Promontory Press, New York, 1971)

Osborn, Fairfield, *Our Plundered Planet* (Little, Brown, Boston, 1948)

Pearson, T. Gilbert, *Birds of America* (Garden City Publ. Co., Garden City, N.Y., 1936)

Stefferund, Alfred, ed., *Birds in Our Lives* (Arco, New York, 1970)

Walker, Lewis Wayne, *The Book of Owls* (Alfred A. Knopf, New York, 1974)

Waters, Frank, *Book of the Hopi* (Ballantine Books, New York, 1963)

Williams, Lt. Col. J. H., *Elephant Bill* (Doubleday, Garden City, New York, 1950)

Wood, Casey A., and F. Marjorie Fyfe, *The Art of Falconry of Frederick II* (Stanford University Press, Stanford, Cal., 1943)

Zimmerman, David R., *To Save a Bird in Peril* (Coward, McCann & Geoghegan, New York, 1975)

Articles

Austin, G. T., and W. G. Bradley, "Additional Responses of the Poor-Will to Low Temperatures" (*Auk,* Oct. 1969)

Brodkorb, Pierce, "A Giant Flightless Bird from the Pleistocene of Florida" (*Auk*, April 1963)

Buscemi, Doreen, "The Last American Parakeet" (*Natural History*, April 1978)

Calder, William A. III, "Energy Crisis of the Hummingbird" (*Natural History*, May 1976)

Dawson, W. R., V. H. Shoemaker, H. B. Tordoff, and Arieh Borut, "Observations on Metabolism of Sodium Chloride in the Red Crossbill" (*Auk*, Oct. 1965)

Domalain, Jean-Yves, "Confessions of an Animal Trafficker" (*Natural History*, May 1977)

Drury, William H., "Abundant Birds of Beringia" (*Natural History*, Feb. 1978)

Gottlieb, Gilbert, "The Call of the Duck" (*Natural History*, Oct. 1977)

Gould, Stephen J., "Were Dinosaurs Dumb?" (*Natural History*, May 1978)

Graham, Frank Jr., "Endangered Birds: Tinkering for Time" (*Audubon*, Nov. 1977)

Kilham, Lawrence, "Reproductive Behavior of the Red-Breasted Nuthatch: I—Courtship" (*Auk*, July 1973)

Laycock, George, "The Pelican No One Knows" (*Audubon*, Jan. 1979)

LeMaho, Yvon, "The Emperor Penguin: A Strategy to Live and Breed in Cold" (*American Scientist*, Nov.-Dec. 1977)

Martin, Calvin, "The War between Indians and Animals" (*Natural History*, June-July 1978)

Moynihan, M., "Display Patterns of Tropical American 'Nine-Primaried' Songbirds" (*Auk*, April 1963)

Odum, Eugene P., and Clyde E. Connell, "Lipid Levels in Migrating Birds" (*Science*, 18 May, 1956)

Ostrom, John H., "A New Look at Dinosaurs" (*National Geographic*, Aug. 1978)

Prichard, C. H., "Endangered Species: Florida Crocodile" (*The Florida Naturalist*, Oct. 1977)

Ricciuti, Edward, "The Bird Lovers" (*Audubon*, Sept. 1977)

Sage, Bryan, "Flare Up Over the North Sea" (*New Scientist* [London], Feb. 1979)

Schmidt-Nielsen, Knut, "How Birds Breathe" (*Scientific American,* Dec. 1971)

Schueler, Donald G., "Incident at Eagle Ranch" (*Audubon,* May 1978)

Short, Lester, "Woodpeckers Without Woods" (*Natural History,* March 1971)

Whiten, A., "Operant Study of Sun Altitude and Pigeon Navigation" (*Nature* [England], June 1972)

Willoughby, Ernest J., "Drinking Responses of the Red Crossbill (*Loxia curvirostra*) to Solutions of NaCl, $MgCl_2$, and $CaCl_2$" (*Auk,* Oct. 1971)

Wolf, Ray, "How Bugs Tune in to Your Garden" (*Organic Gardening,* March 1979)

Young, Howard, Andrew Hulsey, and Robert Moe, "Effects of Certain Cotton Insecticides on Mourning Doves" (*Proc. Arkansas Academy of Science,* 1952)

Zimmerman, David R., "A Technique Called Cross-Fostering May Help Save the Whooping Crane" (*Smithsonian,* Sept. 1978)

INDEX